CURRENT AFRICAN ISSUES 57

Current Status of Agriculture and Future Challenges in Sudan

Farida Mahgoub

NORDISKA AFRIKAINSTITUTET, UPPSALA 2014

INDEXING TERMS:
Sudan
Agricultural projects
Water resources
Water management
Irrigation
Agricultural development
Food security
Climate change

The opinions expressed in this volume are those of the author
and do not necessarily reflect the views of the Nordic Africa Institute.

ISSN 0280-2171
ISBN 978-91-7106-748-7
Language editing: James Middleton
© The author and the Nordic Africa Institute
Production: Byrå4
Print on demand, Lightning Source UK Ltd.

Contents

Part 1
1-1: Abstract 7
1-2: Introduction 7
1-3: General country information 8
1-4: Historic overview of agriculture in Sudan 12

Part 2: Agriculture in Sudan 19
2-1: Current Status 19
2-2: Irrigation 22
2-3: Water resources 26

Part 3: Agricultural schemes in Sudan 34
3-1: Major agricultural schemes 34
3-2: Irrigation schemes 35
3-3: The Gezira Scheme 35
3-4: New Halfa Irrigation Scheme 42

Part 4: Food 48
4-1: World food situation 48
4-2: Food security 49
4-3: Food sovereignty 52
4-4: Food situation in Sudan 54
4-4: Causes of food insecurity in Sudan 59

Part 5: Climate 61
5-1: Agriculture and global food security under climate change 61
5-2: Impacts of climate change in Africa 63
5-3: Impacts of climate change in Sudan 64
5-4: Climate change adaptation measures 72
5-5: Conclusion 73

References 76

Appendix 85

Table of Figures

Table 1 Land use in Sudan 13
Table 2 Real GDP, agricultural GDP and per capita GDP growth rates, 1986–2004 (%) 14
Table 3 Average shares of main agricultural sub-sectors in area and GDP, 2004–06 ... 15
Table 4 Comparison of area, yield and production forecast by crop and region 16
Table 5 Ecological zones of Sudan 25
Table 6 Water supply from the Nile and its tributaries 29
Table 7 Summary of the available water to Sudan 29
Table 8 Main storage reservoir capacities (km3) 30
Table 9 Sudan: Total water requirements, 1957 32
Table 10 Irrigated area of the Gezira Scheme 37
Table 11 Composition of New Halfa Production Scheme by land use 44
Table 12 Number of undernourished people by region, 1990–92 and 2010–12 51
Table 13 Proportion of food insecure in Africa 52
Table 14 Severity of undernourishment (FAO) 55
Table 15 Extreme weather and climate events in Sudan 67

Preface

The theme of this issue is the core of Sudan life and the main driving force for its economy. Despite its importance, the agricultural sector, as income generating sector for 60-80 percent of households and represents about 80 percent of the country's export etc., it has always been neglected and did not get what it deserve of state's care. Fluctuating weather condition is another factor that affects the agricultural sector significantly. The situation had been exacerbated when oil was discovered as all attention had been directed to invest in oil and related industry whereas the agricultural sector had been completely ignored and neglected.

Because of the very small stake of agriculture in the country's budget, the situation is that very rare - if not at all - researches are (apart from that conducted by international organization such as FAO and others) currently taking place in Sudan. Hence, availability of up-to-date empirical data and statistics is very much challenging. In addition, these empirical data and statistics are divergent and lack consistency most of the time.

What have been mentioned so far do not reflect the reality of current situation of agriculture in Sudan as it is more complicated and many factors have played and still play different role besides a great deal of details. Thus, it is a very difficult task to write about agriculture in Sudan if not challenging. In order to write this book I have tried very hard to refer to the most up-to-date empirical data and statistics that available in journals, papers, and internet. The aim is to collect and put some of what have been written by scholars and researchers about different agricultural aspects together in one place. By doing that I tried to give as accurate image as possible about the current status of agriculture in Sudan especially in relation to water and food situation.

The importance of the image I tried to give by writing this book is that: there is a new trend regarding agriculture in Sudan. Large investments in agriculture by acquiring large tracts of most fertile agricultural lands by international companies that belong to China, India, Egypt, Saudi Arabia, Gulf and other countries which have interest to invest in agriculture in Sudan. The same activity is practiced by some national investors. The valid argument is the validity of such activity in a country like Sudan where the state failed to feed its own people and considered as a net importer of wheat. In addition Sudan has a limited share of Nile water according to 1959 Nile Water Agreement. However, the theme of large-scale acquisition of agricultural lands have not been included in this book it rather will be the theme of coming studies.

I have tried to write this book as objectively as possible especially on issues that have political implications such as Nile Water Agreement and its implications on Nubians and other indigenous people who have been affected to a great deal by such an agreement. Any errors of facts and the point of view expressed in this book are solely my own. However, I would like to thank Dr. Terje Oestigaard for hosting me in Nordic Africa Institute, and for discussions and support, and the Nordic Africa Institute for providing me a stimulating and workplace. I would also like to thank Dr. Ali Ayoub, Dr. Salman M.A. Salman, Prof. Ahmed Faris, Dr. Ali Abdelaziz for discussion and good comments. I would also like to thank the following people for provding me different kind of support: my uncle Fikiri Hassan, Dr. Ijaimi Abdalatif and Al-Sadig Hasab Al-Rasool for providing some materials from Sudan; Johan Sävström for coordinating the practical aspects; Linda Karlsson and Huddinge Municipality for supporting economically. Finally, Fadia, Inga-Britt and Abdelhadi for supporting socially, and my husband Mohamed Subahi for his support and feedback.

Farida Mahgoub, Uppsala, January 2014

Part One:

1-1: Abstract

Urbanisation and long-lasting civil wars and conflict mean that the demographic pattern in Sudan is changing drastically. Nevertheless, 60%-80% of Sudanese engage in subsistence agriculture. Agriculture remains a crucial sector in the economy as a major source of raw materials, food and foreign exchange. It employs the majority of the labour force, and serves as a potential vehicle for diversifying the economy. However, no rigorous studies have explained productivity in this sector in relation to food security.

The literature reveals the pervasive inefficiency of Sudanese farmers and large-scale state-owned schemes, such as Gezira Scheme, which produce significantly below their thresholds. Many studies have found that their output levels are less than optimal. This is because of recurrent drought, land degradation, inefficient irrigation infrastructure and inconsistent agricultural policies. The literature also shows that fluctuations in agricultural productivity happen because of fluctuating weather patterns.

The situation has worsened because agriculture in particular has been neglected since the advent of oil production in the early 2000s. Moreover, Sudan's agricultural growth has been unbalanced, with the majority of irrigated agriculture concentrated in the Centre and a huge disparity in development indicators between the best- and worst-performing regions. Thus, studies show that the vast majority of Sudanese are reported to be food insecure, especially internally displaced persons (IDPs) and in conflict regions such as Darfur, Kordofan and other regions.

1-2: Introduction

After decades of civil conflict and associated political instability, because of human-induced and recurrent natural disasters (floods, droughts, outbreaks of livestock diseases), millions of people in Sudan continue to face severe and chronic food insecurity. Given that between 60 per cent and 80 per cent of the working-age population rely on agriculture for their food and livelihoods, the sector's importance to economic recovery and the consolidation of long-lasting peace in Sudan cannot be ignored. At the same time, the new phenomenon of large- scale land acquisitions of agricultural land is taking place in Sudan. Many countries such as China, Saudi Arabia, Gulf countries have interest to invest in agricultural lands. However, only one case of large- scale land acquisition is mentioned in this report. This case is mentioned in relation to The Gezira Scheme and its deterioration. I argue over the validity of the practice in Sudan, backed by images of the poor, peasants, pastoralists and IDPs displaced by conflicts and environmental crises.

I have conducted a thorough desk review using a large body of literature on national and local knowledge on agriculture in Sudan. The purpose was to bring together several studies on agriculture and provide a platform for coming studies on the mentioned phenomenon of large- scale land acquisitions of agricultural land, and thereby equip observers and readers with a holistic picture of agriculture in Sudan. This will allow them to understand the potential scale of the implications that the phenomenon poses to agriculture and the people – especially those who depend on different land uses for their food and livelihoods – in terms of national food and water supplies. This will be achieved by shedding light on the current status of agriculture and water resources, as well as the food situation in Sudan, especially with regard to recurrent droughts, desertification and climate change conditions.

The report consists of five parts. The first part comprises the introduction and general country information; it ends with historical background about agriculture in Sudan, and how agricultural production has fluctuated according to changes in weather patterns. The second part reflects on the current status of agriculture in the country, and various water resources and irrigation methods. The third part sheds light on major agricultural schemes and refers to the examples of the Gezira and New Halfa Schemes. The food situation globally, regionally and locally makes up part four. The picture would not be complete without taking into account the extent of impact of climate change on agriculture in general and water resources in particular and consequently on food security. That is the theme of part five, which ends by suggesting a number of climate change adaption measures.

The study draws on a literature review; qualitative interviews with major Sudanese scholars; and electronic national and international newspapers in English and Arabic.

1-3: General country information[1]

Sudan occupies a region that is located in the middle part of the Nile Basin to the south of Egypt. The country is located within the Sudano-Sahelian region (Frenken 2005) in north east Africa, with geographic coordinates: 4° and 22° north and longitudes 22° and 38° E (Zaroug 2006), and has a special geopolitical location that bridges the Arab world and sub-Saharan Africa: it facilitates trade and human movement between, and is a melting pot of, African and Arab cultures. Sudan achieved independence in 1956. The country comprises four regions divided into 15 states. Its total area was reduced from 2,500,000 km² to 1,882,000 km² (Frenken 2005) following the independence of South Sudan in

---

1. The focus in this study is on Sudan before the secession of Southern Sudan.

2011. Together, the two countries contain 63 per cent of the Nile basin (Shinn 2006) and share borders with nine countries.

Sudan is divided into five distinct ecological zones: desert, semi-desert, woodland savannah, flood region and montane vegetation. The most important crop species, especially during times of drought, are the indigenous fruit *kursan* and the vegetable okra (Bashir 2001). The country is traversed by the Blue Nile and White Nile rivers, which meet in the capital Khartoum to form the main Nile River, which flows north into the Mediterranean Sea. The two Niles and their tributaries have varying degrees of influence on irrigated agriculture and livestock production systems. There are also a large number of seasonal rivers and water courses known as *wadis*, of which Gash (Mareb) and Baraka are the largest.

Sudan has constructed five dams across some of these bodies of water. The Roseires and Sinnar dams are on the Blue Nile; Khashm el Girba on the seasonal Atbara river; Jabal Awliaa on the White Nile; and Marawe on the Nile. These dams provide water for irrigation, fishing and to generate electricity. Erratic rainfall and recurrent spells of drought emphasize the importance of reliable sources of groundwater to rural areas, as well as remote urban centres. The water bearing rock strata comprise the Nubian Sandstone, the Um Rwaba Series and the basement complex (Abdalla and Karar 2010). Although Sudan lies within the tropics, the climate ranges from hot and dry to arid desert, with a rainy season between April to November that varies by region (Central Intelligence Agency 2013).

Soils in Sudan can geographically be divided into four categories: sandy in the northern and west-central areas; clay in the central region; and laterite in the south; with alluvial soils as a fourth, less extensive and widely separated category. Alluvial soils have great economic importance. They are found along the main Nile to Lake Nubia; in the delta of the Gash River in the Kassala area; in the Baraka Delta in the area of Tokar near the Red Sea in Ash Sharqi State; and along the lower reaches of the White Nile and the Blue Nile rivers (Central Intelligence Agency 1991).

Clays in central Sudan are agriculturally the most important soils, and extend from southern Kordofan through Al-Awsat and west of Kassala. They are characterised by cracks when they dry out during the dry months, which allows restoration of soil permeability. These soils are used in many schemes: either irrigated ones such as the Gezira, Rahad, El Suki, New Halfa, and the Blue Nile and White Nile Schemes; or the mechanised, rain-fed schemes in Gedaref, Sennar, Blue Nile and South Kordofan States. Traditional cultivators also use these soils in all rain-fed areas where they grow sorghum, sesame, peanuts, and cotton (Central Intelligence Agency 1991).

The cultivable area ranges from 84 million hectares (ha) to 105 million ha

Courtesy of the University of Texas Libraries, The University of Texas at Austin
Map 1: Former Republic of the Sudan

(UNEP 2007), with reasonably fertile soils (Government of Sudan 2008). The terrain is generally flat; featureless plain and desert dominates the north. Eight national parks cover a total area of about 8.5 million ha, representing 3.2% of the country's area. Two national parks have been designated as "biosphere reserves"; one is a marine park on the Red Sea coast. There are also 11 game reserves (3.3 million ha), which constitute 1.3% of Sudan's total area, and three wildlife sanctuaries (95,000 ha). The total area of protected land is about 11.9 million ha, or 5.4% of the country's total surface area, but the conservation status of all protected areas is rated as unsatisfactory (Bashir 2001).

Courtesy of the University of Texas Libraries, The University of Texas at Austin

Map 2: The New Republic of the Sudan

The population in 2012 was 37,195,349 (World Bank 2012), which is growing at the rate of 2.53 per cent (Government of Sudan 2008). The average population density is about 10 people per km^2, and 63 people per km^2 on arable land. In contrast, the population density of cultivated land is approximately 370 people per km^2. Much of the population is clustered in central Sudan and along the Nile River and its tributaries. Sudan is multi-cultural, multi-ethnic, multi-lingual and multi-religious (Shinn 2006).

For decades, Sudan was embroiled in two civil wars that ended in 2005. The first took place in 1955-72 and broke out again in 1983. Peace talks began in 2002. The second war had famine-related effects, which displaced more than 5 million people and; more than 2.5 million people died (Watch 2011). Most commentators have attributed the country's political and civil strife either to an age-old racial and ethnic divide between Arabs and Africans or to colonially constructed inequalities (Johnson 2003).

In January 2005, the final North/South Comprehensive Peace Agreement (CPA) was signed, and on 9 July 2011 South Sudan seceded and became independent. Conflicts over water and grazing rights in the western region of Darfur and elsewhere have become entwined with political rivalries on a larger scale, including in neighbouring countries. Conflict broke out in Darfur in 2003 that was estimated to have caused between 200,000 and 400,000 deaths and displaced nearly 2 million people.

The struggle has become increasingly regional in scope and has brought instability to eastern Chad. Sudan has also faced large refugee influxes from neighbouring countries primarily, from Ethiopia, Eritrea and Chad, with the attendant problems of poverty and food insecurity (UNICEF 2004).

1-4: Historic overview of agriculture in Sudan

Agriculture in Sudan is the principal source of income and livelihood for between 60 per cent and 80 per cent of the population (Elgali, Mustafa et al. 2010), and the engine of growth for other economic sectors such as trade, industry and transport. The expected results of agricultural development would be to create more job opportunities. This would make rural areas more habitable and reduce internal migration to big cities, which would lead to stable food security status and poverty reduction.

Moreover, a sound agricultural base would balance distribution of the benefits of development between the different states and localities by giving more attention to the least developed ones (Government of Sudan 2008). This does not currently apply to the agriculture sector in Sudan. It has failed to fulfil such objectives, which has affected the food and water situation in the country, as will be explained further.

Sudan is endowed with large areas of cultivable land, which are situated between the Blue Nile and the White Nile, and in the region between the Blue Nile and the Atbara river. Other regions with cultivable land are the valleys of the plains, where irrigation is extensively used, and in the narrow Nile valley. This land has different uses, as illustrated in Table 1. Arable land constitutes approximately one-third of total area of the country, of which 21 per cent is cultivated with fluctuating productivity – but output remains far below potential performance.

More than 40 per cent of the total area consists of pasture and forests (Stads and ElSiddig 2010). Subsistence farming and commercial production for local consumption and export are practised. Five main types of farming exist in Sudan (UNEP 2007):

- Mechanised rain-fed agricultural schemes
- Traditional rain-fed agriculture
- Mechanised irrigation schemes
- Traditional irrigation
- Livestock husbandry/pastoralism

Livestock production is mainly based on traditional pastoral systems; 90 per cent of livestock in the country belong to traditional pastoral production systems (Zaroug 2006). However, pastoralists are now opting to settle, becoming agro-pastoralists (Ali Ayoub).

Table 1 Land use in Sudan

Item	Area ('000 ha)
Total area	250,429
Land area	237,443
Area under water	12,986
Arable land	84,034
Cultivated land	17,471
Uncultivated land	66,563
Forest and wood land	64,360
Other	49,569

Source: Administration of Statistics and Information (1995)

Since the time of the Anglo-Egyptian Condominium (1899–1955), the economy has relied heavily on cotton exports (Abbadi and Ahmed 2006). Cotton cultivation used rotational cropping, with a long fallow period to avoid the use of chemical fertilisers, until the 1970s when intensification and diversification were introduced to increase output (Ali Ayoub). According to FAO, the country's agricultural policy was changed in 1997 in an effort to attain greater food self-sufficiency. It focused on reducing the area of cotton in production because of a shortage of irrigation water; and instead replaced the crop with wheat and sorghum, which require less water.

The hope was that, regardless of the inevitable reduction in export revenue, any increases in food production in the long run would mitigate the increasing effects of drought and food shortage (FAO/EWSFA 1997) and reduce cereal imports. Mechanised rain-fed agriculture also expanded. In addition to land allocation policies, this led to the displacement of subsistence farmers and nomads, and dismantled traditional systems of communal ownership and management (IFAD 2010).

Furthermore, agriculture was a low government priority; the meagre budget allocation for agriculture dwindled from 3.4% to 1.6% during 2000–05 (Hamid Faki 2012). This top-down approach to development reduced rural producers to 'policytakers' rather policymakers. Lack of political stability, and weak government administrative and implementation capacity, have been contributory factors to the decline of county's growth rate (Government of Sudan 2008).

In addition to man-made factors, fluctuating weather patterns led to a decline in agricultural production. Average annual growth to 0.8 per cent for the period 1980–87, compared with 2.9 percent for 1965–80. Consequently, the agricultural sector's total contribution to GDP has declined over the years, even as the other sectors of the economy have expanded: it had fallen to about 36 percent in 1988 from 40 percent in the early 1970s (Mongabay 1991).

Another study showed a decline of 28% during mid-1980s and early 1990s (Abbadi and Ahmed 2006). Table 2 compares the growth rate of agricultural GDP to growth rates of other macroeconomic measures between 1986 and 2004. During the period from 1986 to 1999 (Table 2), the average growth rate of agricultural GDP went from −1.2 per cent to 10.9 per cent. This increase was attributed to favourable weather conditions, which revived the sector (Abbadi and Ahmed 2006). High investment in the oil sector and related services caused a decline in the growth rate of 5.58% between 2000 and 2004, a negative effect of government planning and policies (Stads and ElSiddig 2010).

A further study attributes the decline in 2003 mainly to deterioration in traditional rain-fed agriculture; and the further decline in 2004 to the poor performance of mechanised rain-fed agriculture (Shukri Ahmed, Getachew Diriba et al. 2007). However, the growth rate increased to 7% in 2005. The agricultural sector continued its dominance, accounting for 39 per cent of GDP compared with 34 per cent for services and 28 per cent for manufacturing (Shukri Ahmed, Getachew Diriba et al. 2007); but in 2007 agriculture's share dropped to 33%. A third study by Siddig et al. (2012) showed that the decline in the contribution of agriculture to total GDP is only in percentage terms and that the sector's GDP value has been increasing. The study explained that the sector is not deteriorating but other sectors are growing faster. For example, that industrial sector's share of total GDP grew from 15 per cent in 1997 to 31 per cent in 2008. The share of the services sector increased from an average of 33 per cent over the previous ten years to 34 per cent in 2007, and 33 per cent in 2008 (Siddig and Babiker 2012).

The average area under crops during the period 2004–06 slightly decreased to less than 17 million ha (Table 3). The distribution of crop cultivation over

Table 2 Real GDP, agricultural GDP and per capita GDP growth rates, 1986–2004 (%)

	1986–90	1991–95	1996–99	2000	2001	2002	2003	2004
Growth rate of real GDP	2.3	3.5	5.7	5.1	8.0	6.5	8.3	7.2
Growth rate of agricultural GDP	−1.2	0.7	10.9	5.3	5.6	7.3	5.2	4.5
Growth rate of per capita GDP	−0.8	1.6	3.6	2.8	5.6	4.2	2.2	1.5
Inflation	43.3	106.4	43.6	8.0	4.9	8.3	7.4	8.7

Source: World Bank (2003), Bank of Sudan (2000–04) (Abbadi and Ahmed 2006)

Table 3 Average shares of main agricultural sub-sectors in area and GDP, 2004–06

Sub-sector	Area (million ha)	Share of area (%)	Share of GDP (%)
Irrigated (two sub-sectors)	1537	9	11.3
Traditional rain-fed	9812	58	6.3
Mechanised rain-fed	5513	33	1.4
Total	16,861	100	19.0

Source: Calculated from the Economic Survey 2006, Ministry of Finance and National Economy (Hamid Faki 2012)

the main farming types indicated the wide prevalence of traditional agriculture (58%) and sizeable mechanised cropping (33%). Irrigated farming, though smaller in area, was very important in terms of total value of production and contribution to the country's GDP relative to other two sectors (Table 3) (Hamid Faki 2012).

Cereals dominate crop production in Sudan and provide nearly 53% of the population's daily calorie requirements (FAO-SIFSIA. 2010). The same high levels of annual fluctuation as for other crops characterises the production of the major staples – sorghum, millet and wheat. Before 1960, apart from small areas in Darfur and Kordofan, Sudan grew wheat only in the northern state, and even there only on limited scale.

Although environmental and climatic conditions are less favourable for wheat than in the north, the government decided to grow wheat on the Gezira Scheme, between the White and Blue Niles south of Khartoum, because of a local land shortage and high cost of irrigation water in the north. At the same time, wheat cultivation was extended to the New Halfa Agricultural Production Scheme in the east, on the Atbara River (FAO 2000).

During the agricultural seasons in 2000–05, about 1.89 million ha of arable land were under irrigated agriculture, 8.37 million ha under traditional rain-fed cultivation and 5.44 million ha under mechanised farming (Shukri Ahmed, Getachew Diriba et al. 2007). In 2006, more areas were added to all these three sectors because of favourable rains. Despite a few outbreaks of pests or diseases, the 2006–07 season produced a record cereals harvest of 6.64 million metric tons. These yields were 22 per cent higher than in 2005, and production across all three sectors was considerably improved: 36 per cent higher than the previous year's average and above the long-term average (Shukri Ahmed, Getachew Diriba et al. 2007).

The production of cereals, sorghum, millet and wheat declined in 2010 by nearly 42% from an average of 4.9 million metric tons in 2006-09 to only 2.9 million metric tons. The magnitude of production decline varied by crop: 46.8 per cent for sorghum; 31.3 per cent for wheat; and 24.2 per cent for millet. The largest production decrease occurred in the mechanised rain-fed farms, which contributed nearly 32 per cent of national cereals output (FAO-SIFSIA. 2010).

Table 4. Comparison of area, yield and production forecast by crop and region

Region	Harvested area ('000 ha)						Yield (metric tons/ha)						Production ('000 metric tons)					
	01–02	02–03	03–04	04–05	05–06	06–07	01–02	02–03	03–04	04–05	05–06	06–07	01–02	02–03	03–04	04–05	05–06	06–07
Sorghum																		
Northern	171	70	120	48	72	184	2.16	1.73	1.31	1.72	1.75	1.36	369	121	157	83	79	249
Central	1749	1256	2208	960	2009	1940	0.99	0.83	0.94	0.95	0.73	1.02	1732	1010	2065	910	1431	1984
Eastern	1407	1429	2365	999	1896	1948	0.49	0.48	0.64	0.53	0.55	0.58	687	691	1509	533	1090	1132
Kordofan	1046	1026	971	799	990	1296	0.50	0.36	0.38	0.45	0.55	0.56	528	365	366	362	546	726
Darfur	753	591	448	224	341	334	0.64	0.41	0.58	0.46	0.65	0.73	480	241	260	102	223	246
South	799	631	969	789	1138	1057	0.84	0.80	0.84	0.90	0.80	0.82	673	503	824	714	906	866
Sub-total	5925	5003	7081	3819	6446	6759	0.77	0.61	0.73	0.71	0.66	0.41	4469	2931	5181	2704	4275	5203
Millet																		
Northern	–	–	–	–	–	–	–	–	–	–	–	–	–	–	–	–	–	–
Central	84	91	180	124	148	227	0.30	0.35	0.42	0.17	0.43	0.51	25	33	75	21	61	115
Eastern	32	23	160	21	53	63	0.47	0.39	0.61	0.31	0.30	0.40	15	9	97	6	17	25
Kordofan	1146	863	1049	488	1076	1082	0.15	0.19	0.18	0.10	0.18	0.23	177	165	189	50	199	246
Darfur	1660	1460	1182	652	904	769	0.22	0.28	0.36	0.31	0.41	0.41	363	374	423	200	370	312
South	8	0	0	0	62	115	1.25	0	0	0	1.58	0.83	10	0	0	3	98	95
Sub-total	2930	2437	2570	1285	2243	2256	0.20	0.25	0.30	0.22	0.33	0.35	590	581	784	281	745	792
Wheat																		
Northern	60	67	77	76	90	103	2.70	2.94	2.40	2.84	2.93	3.46	162	197	185	215	263	354
Central	38	37	82	101	74	145	1.74	4.30	1.88	2.13	1.96	1.95	66	159	154	216	146	284
Eastern	2	2	8	2	3	2	7.50	2.00	1.75	1.48	1.67	1.98	15	4	14	3	5	4
Kordofan	–	–	–	–	–	–	–	–	–	–	–	–	–	–	–	–	–	–
Darfur	3	3	2	1	2	0	1.00	1.00	1.33	1.00	0.5	1.24	4	4	2	1	1	1
South	–	–	–	–	–	–	–	–	–	–	–	–	–	–	–	–	–	–
Sub-total	103	109	169	180	169	250	2.40	3.34	2.11	2.42	2.46	2.57	247	364	356	435	415	642
COUNTRY TOTAL	8958	7549	9821	5282	8689	9015	–	–	–	–	–	–	5306	3876	6328	3420	5435	6637

Source: Shukri Ahmed, Getachew Diriba et al. 2007

Table 4 compares regional figures for 2006–07 with the figures for the preceding five years from 2001–06.

Figure 1 illustrates how wheat imports started to increase since 1990s up to 2006, which reflected the change in the population's food consumption patterns. Sorghum production fluctuates because it is mainly grown in traditional and semi-mechanised rain-fed areas (75% of total sorghum production) (Abbadi and Ahmed 2006). Nevertheless, there were disparities in grain production across various agricultural sub-sectors during the past two decades.

Crop production from traditional rain-fed farming has grown since the early 1990s; it has surpassed the level of semi-mechanised farming, which shrank during the same period. Semi-mechanised system has ceased to be the dominant source of food (sorghum) for Sudan (Institute for Security Studies 2005). However, the contribution of the irrigated sector has remained relatively

Figure 1. Sudan – Imports of wheat and production of sorghum, 1990–2006 ('000 tons)

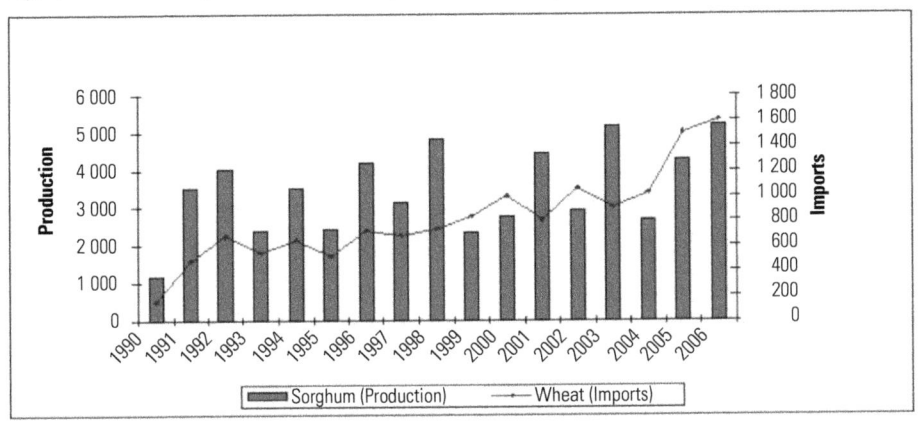

Source: (Shukri Ahmed, Getachew Diriba et al. 2007)

stagnant, apart from its surge in production in response to drought and locusts attacks in 2001–02, and again in 2006–07 when wheat prices increased. However, this production system clearly has the capacity to make a major contribution to food production as a result of increased harvested area (Institute for Security Studies 2005).

Sudan's total cereals production is usually sufficient to meet domestic needs, especially in terms of sorghum and millet, but is a net importer of wheat (Ahmed 2010). Generally speaking, in terms of availability of arable land and different water resources, the country has the potential to become the main food provider for Africa and the Middle East. Over the past few decades, however, variability of rain, seasons of severe drought, problems with food distribution and civil war, and above all mismanagement and lack of knowledge have left the country with recurring food shortages (Ahmed 2010).

Sudan's agricultural exports can be divided into three categories: (i) field crops (e.g. cotton, sesame, peanuts, sugar); (ii) livestock (e.g. sheep, camels and cattle); and (iii) gum Arabic, which represents the major forest exports (Mohamed Alameen 2009). These exports were the main source of foreign currency until the late 1990s when oil replaced them. From that time until the secession of South Sudan, the country turned from an agricultural to a petroleum exporter, following the unprecedented boom of its petroleum export revenues.

The trade balance for the fiscal year ending September 2000 achieved a remarkable improvement, with a surplus of US$226.2 million that was directly attributed to the introduction of oil exports, as well as the reform of economic policies geared towards encouraging exports (Institute for Security Studies 2005). Meanwhile, agriculture showed a dramatic deterioration in its contribution to the country's exports, falling to 8% in 2006 and to 3% in 2007, down from an average of 74% in the 1996-98 period (Ahmed, Sulaiman et al. 2012). Both the relative share and absolute value of agricultural exports have declined (Cesar Guvele 2011).

Agriculture has a significant role to play in the country's development, in terms of exports as well as industrialisation – for example, as an incubator for major manufacturing industries such as edible oils, leather, and sugar (Elgali, Mustafa et al. 2010). Nevertheless, it remains the cause of the country's most serious environmental problems. These include: (i) land degradation (e.g. riverbank erosion); (ii) the emergence of invasive species; (iii) use and mismanagement of pesticides and other agro-chemicals; (iv) water pollution (UNEP 2007); (v) the spread of malaria; and (v) the introduction of perfect conditions for water-borne diseases such as bilharzia (Ali Ayoub).

Part 2: Agriculture in Sudan

2-1: Current Status

As mentioned above, Sudan's primary resources are agricultural, but oil production and exports grew in importance between October 2000 and the secession of the South in 2011. The contribution of agriculture (agro-industry) to total manufacturing output is 60 per cent, in the form of raw materials; 80 per cent of non-petroleum exports are agricultural products (IFAD 2009).

Livestock are raised mainly by pastoral and agro-pastoral groups (who also rely on cultivation). Herd size may vary from below 50 head of cattle to several thousand per household. Pastoral herds are mainly semi-nomadic, as practised in western Sudan and along the southern Blue Nile where traditional herd movements occur between wet- and dry-season grazing areas. The rainy season affords rich grazing, and provides pasture and water. However, mud and biting insects experienced during the rainy season in other areas disturb the herds (Babiker 1997).

In the Butana region (central eastern Sudan) household economies are based on an agro-pastoralist system of production, where livestock (goats, sheep, cattle and camels) and crop production (sorghum) are practised (Babiker 1997). The location within this important grazing area and pastoralist mobility patterns determine the proportion of each type of animal in the herd, as well as the relative importance of livestock production and cultivation to each household (Babiker 1997). This differs from pastoral production systems in western Sudan where tribes are specialised camel owners (*Abbala*) or cattle owners (*Baggara*), and limited cultivation is practised to meet all or part of the household grain requirement. Livestock production has vast potential and many animals, particularly camels and sheep, are exported to Egypt, Saudi Arabia, and other Arab countries. Beef has lately been exported to these countries as well (US Department of State 2012).

Sudan used to depend mainly on traditional agricultural exports, which came from irrigated, rain-fed traditional and livestock sub-sectors. But it exports sugar, molasses, hide, chrome and gold (Ahmed 2010). These manufacturing and mining commodities are a small share of the country's economy (Ahmed 2010).

Oil exploitation has caused a major shift in the country's economic structure over the past two decades (Ahmed, Sulaiman et al. 2012). This has included low levels of inflation, a high GDP growth rate, and the financing of major public works projects, such as development of satellite and digital communications, building new highways and power generation plants. However, the prospect of quick gains in the service and construction sectors, compared to higher risks and lower returns in agriculture, drove most investment activities and commercial

Figure 2. Contribution of Agricultures Sector to Gross Domestic Product (GDP)*

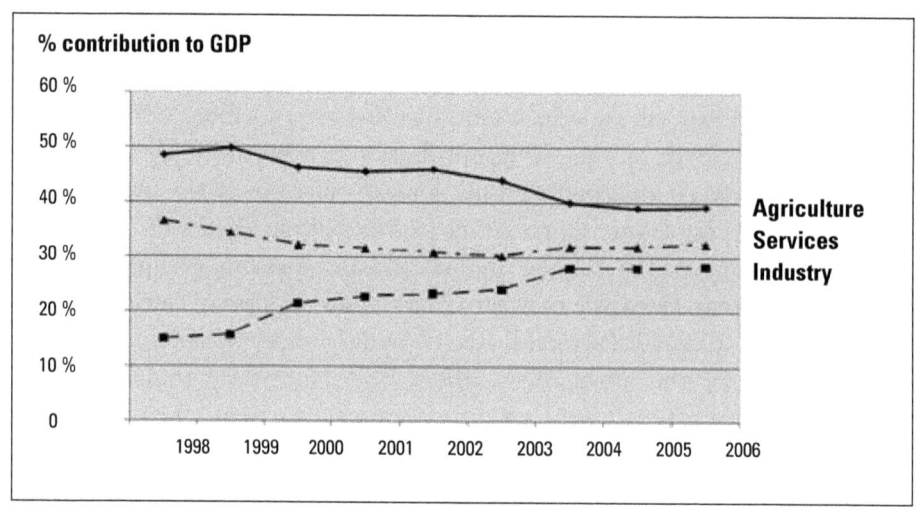

* Factor Cost (constant 1981/82 prices).
Source: (IFAD 2009).

lending in the country, which led to further deterioration of the agricultural sector and its export prospects (Konandreas 2009).

The composition and value of exports had changed radically since the discovery of the oil until the secession of the South of Sudan. And the significant impact of oil on total exports had masked the continued importance of agriculture in Sudan (IFAD 2009). During that period of time, non-oil exports became insignificant in value and as a percentage of total exports, and oil exports predominate, amounting to 74.8% of all Sudanese exports in 2000 and reaching a record high of more than 95% in 2008 (Ahmed 2010). As a consequence of and/or in parallel with the oil boom, consumption patterns of food and durable goods have changed substantially, fuelled also by the return of skilled migrants to work in the oil and construction business; hence, dependence on imported foods have increased considerably (Konandreas 2009).

However, booming oil exploitation has revived the industry sector: its contribution to the economy has almost doubled. The share of the service sector has remained fairly constant, with a small decline because of a reduction in private services (IFAD 2009) (Figure 2).

Sudanese policymakers recognised – or remembered – the importance of agriculture as an export earner and the principal sector for securing and improving the livelihoods of rural people when the country received a shock from soaring food prices caused by the volatility in the oil market in 2008 (Konandreas 2009). In addition, Sudan's accession to the World Trade Organization (WTO) necessitates increasing productivity in the agricultural sector and improving its competitiveness in local and international markets. All of that has underscored the urgent need to emphasise development of the agricultural and industrial sectors (Government of Sudan 2008).

To achieve long-lasting agricultural development, the Government announced its "Green Mobilization" programme in 2006 and adoption of the Strategic Five-Year Plan (2007-11). The Agricultural Revival Plan (ARP) aimed to broaden the base of rural development and export earnings, with particular emphasis on reactivation and diversification of non-oil exports. The High Committee for the Study of the Current Situation in the Agricultural Sector was formed to implement the plan. The methodology the committee adopted to fulfil its terms of reference was as follows (Government of Sudan 2008):

Formation of five specialised sub-committees to prepare comprehensive reports covering all issues specified in the terms of reference. The committees were:
- The committee for diagnosing the current situation in the agricultural sector.
- The committee for analysing the crop mix and requirements of food security and export.
- The committee for increasing productivity.
- The committee for agro-based industrialisation
- The committee concerning policies supporting agricultural development.

According to the Executive Program for Agricultural Revival document, a SWOT analysis was used to diagnose the current situation in the agricultural sector (Government of Sudan 2008). The author argues that this diagnosis did not reflect the reality of the current situation of agriculture in Sudan because it lacked data that could be used as a departure point for any development plans. This applies to all documents that were prepared in relation to implementing the agricultural revival programme. In any event, the programme was not fully implemented and came to a halt[2].

The secession of south Sudan in July 2011 has overshadowed Sudan's macroeconomic development and had a negative impact on the overall life of the Sudanese people. According to African Development Bank (AfDB, OECD et al. 2012), features are:
- Real gross domestic product (GDP) is expected to grow modestly in 2013, mainly because of the loss of oil revenue and decline in population following the secession of South Sudan.
- The government of Sudan has attempted to address heightened economic and social challenges through the introduction of austerity measures.

---
2. To read more about the Agricultural Revival Plan, please refer to the document at: http://www.cmi.no/sudan/doc/?id=1147

- Youth unemployment, particularly among university graduates, is high and increasing.

However, although agriculture continues to provide the majority of export revenue outside of the oil sector, growth in recent years has been tepid (IFAD 2009). Abbadi and Elhag Ahamed (Abbadi and Ahmed 2006) summarised the constraints:

- Reduced competitiveness because of low productivity and high marketing costs, which results in lower prices for farmers.
- Exports of most goods are concentrated in a few foreign markets, which make them vulnerable to disruptions, such as an import ban on sheep sales in Saudi Arabia in 2000 and 2001 and again in 2007.
- Lack of strategic planning for different agricultural sub-sectors.
- The low priority accorded to the sector, which is reflected in allocation of public expenditure (3% of the total for the country), formal banking credit (11%) and investment (3%).
- Inadequate complementarities and coordination of macroeconomic and sector policies, and persistent neglect of the role that small producers play in achieving food security and poverty alleviation.
- Instability of production that exposure to natural risks and hazards causes, in addition to price competition from subsidised imported goods.
- The low productivity of animal and crop producers because of inadequate training, and a lack of extension programmes or supportive producers' associations.
- The inefficient use of human resource capacities engaged in agriculture.
- The meagre budget allocated to agricultural research (0.04%) of public expenditure (Abbadi and Ahmed 2006).
- Inadequate social and physical infrastructure.
- Weakness of laws governing the lease and use of land.

A UNDP study showed that the welfare of the population, especially of the poor who are mainly located in rural areas, drastically depends on the performance of the agricultural sector (UNDP 2006). A more recent study showed that the current situation of agriculture in Sudan is fragile. Nevertheless, enormous potential remains that could enable the country to raise crop yields and bridge at least part of its current 'yield gaps' – the shortfall between actual and potential food production (Cesar Guvele 2011).

2-2: Irrigation

The variety of agricultural zones in Sudan means that the country is suitable for a wide range of crops. Agriculture depends principally on rainfall and irrigation from major rivers – the Nile and its tributaries. Crops are also cultivated under

Map 3. Water courses

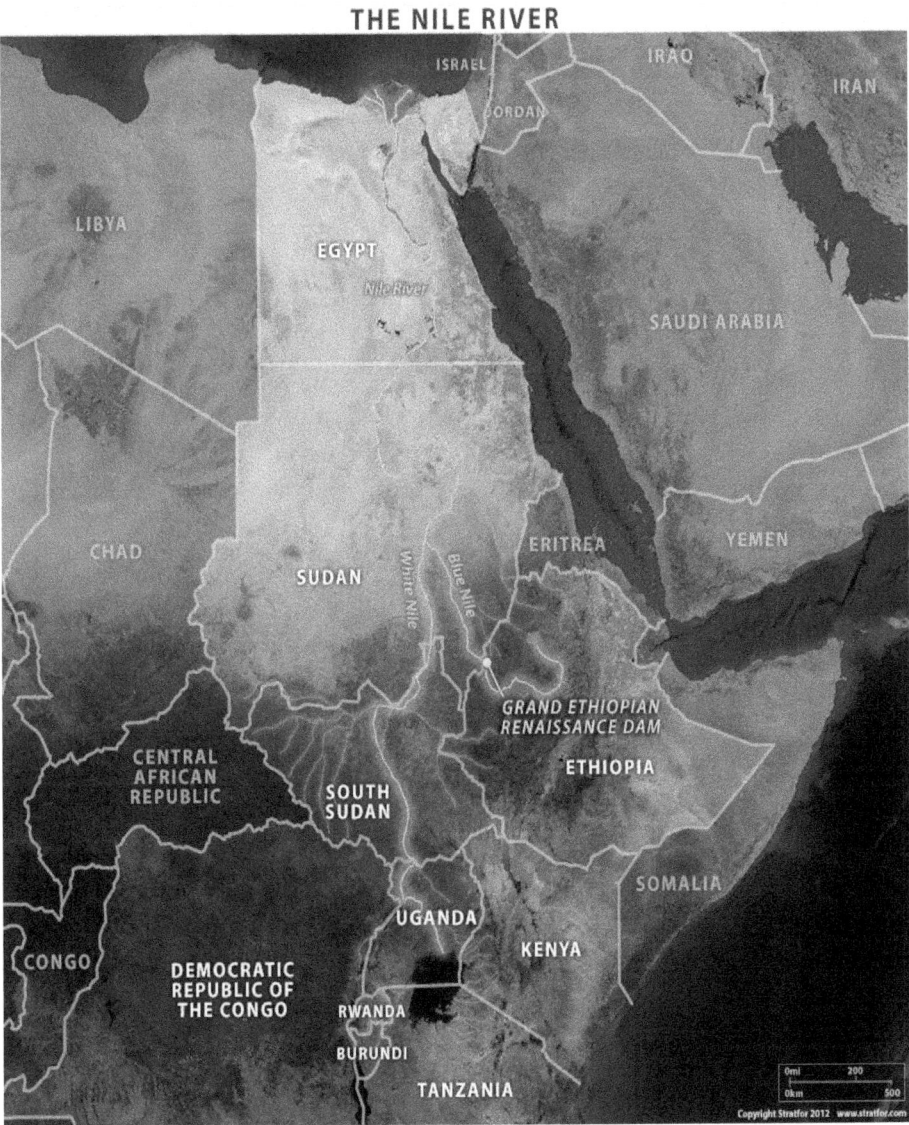

Source: Stratfor at: www.stratfor.com

flood irrigation schemes fed by seasonal rivers, especially in the east and west. For centuries, traditional irrigation has used the *shaduf* (a traditional device to raise water) and *sawagi* (a waterwheel to lift water to irrigate smallholdings) to take advantage of the Nile waters and annual flooding (Metz 1992).

Since the 1970s, some of these devices have been replaced by more efficient diesel pumps, and more recently electric pumps. In general, the water supply for all irrigation is provided by dams and/or large pumps, with extensive networks of canals for irrigation and drainage covering entire schemes (Frenken 2005).

In 1954, projects that could be irrigable by gravity or pumped water in Sudan, in an area that extended for about 400km from north to south and 30km from east to west, were divided as follows:
- Areas in the catchments of the Rahad and Dinder rivers, from the Ethiopian border to their confluence with the Nile.
- Areas between Sennar and the Ethiopian border irrigable from the Blue Nile.
- Areas irrigable from the White Nile between Kosti and the mouth of the Sobat.

Thus, all projects in these areas would be watered from either the Blue or White Niles (Taha 2010). The highly mechanised major schemes within these areas are the Gezira, New Halfa and Rahad Schemes that together make up sub-Saharan Africa's most extensive irrigated area (Holt and Coulter 2011). The Gezira and New Halfa Schemes will be discussed in more detail below.

At independence, canals irrigated approximately 809,371 ha and half of that area was in the Gezira Scheme (Wallach 1988). Gravity flow was the main form of irrigation, as it was in the Gezira Scheme, but pumps served about one-third of the irrigated area (Metz 1992) on the Nile downstream from Khartoum, and on the left bank of the White Nile downstream from Kosti. There were also small irrigated schemes at Tokar on the Red Sea coast and on the inland Gash Delta near Kassala. All these schemes, with the exception of some pump schemes on the main Nile, had been built to generate export revenues from the sale of long-staple cotton, and were of great importance to the economy (Wallach 1988).

In the 1970s, new projects were undertaken: 161,874 ha at New Halfa; and 121,406 ha at Rahad, on the right bank of the Blue Nile, opposite the Gezira Scheme (Wallach 1988). The greater part of this irrigated area (93%) was in government projects and the rest (7%) in the private sector. The Nile and its tributaries remained the main source of water for 93 per cent of irrigated agriculture; 67 per cent came from Blue Nile. In addition to gravity schemes, pump irrigation accounted for about 25 per cent of the irrigated area, which produced some wheat, as well as most of the country's fruit, vegetables, winter legumes and spices (FAO/EWSFA 1997).

Sudan is divided into different ecological zones (Harrison and Jackson 1958).

Table 5 shows these zones and major agricultural enterprises in each zone indicated below:

Desert: in this zone irrigated agriculture exists along the banks of the Nile and Atbara rivers, and on the neighbouring lands where irrigation water is

Table 5 Ecological zones of Sudan

Zone	% of Sudan area	Mean annual rainfall (mm)	Wet season	Dry season	Main land use types
Desert	28.9	<75	July–Sep.	Oct.– June	Irrigated agriculture; grazing along seasonal water courses
Semi-desert	19.6	75–300	July–Sep.; Nov.–Jan.	Nov.–June; March–Sep.	Irrigated agriculture; dry land farming with water harvesting; pastoral
Low-rainfall savannah	27.6	300–800	May–Sep.	Nov.–April	Irrigated agriculture; rain-fed traditional cultivation; mechanised farming; pastoral; forestry
High-rainfall savannah	13.8	800–1500	April–Oct.	Dec.–Feb.	Rain-fed traditional cultivation; mechanised farming; pastoral; forestry
Flood region	9.8	600–1000	May–Oct.	Dec.–April	Traditional cultivation; pastoral; wild life
Mountain vegetation	0.3	300–1000	Variable	Variable	Traditional cultivation; pastoral; forestry; horticulture

Source: (Harrison and Jackson 1958)

conveyed by canals; mainly small privately owned plots, semi-governmental, cooperative agricultural schemes, as well as privately owned schemes growing field crops, vegetables, spices and fruit trees. Different systems of irrigation include:

- Basin irrigation that depends on diversion of Nile waters during flood periods; crops are then grown using residual moisture stored in the soil; crops include wheat, sorghum, fava bean, field bean, maize, lablab, fruits, and vegetables.
- Water pumped from the Nile and conveyed by canals to irrigate fields where crops such as cotton, sorghum, wheat, beans, spices, alfalfa, sorghum, fruit trees and vegetables are grown.
- Water pumped from ground aquifers and conveyed by canals or modern irrigation systems (drip, central, or pivot) to fruit trees, vegetables and field crops.

Semi-desert: rainfall is less than <300 mm/year in this zone; rain-fed cultivation is restricted to traditional farming on the *Qoz* sand (mainly millet). In areas with higher clay content, run-off harvesting is practised to grow sorghum. Irrigated farms that use water from the Blue Nile, White Nile and Atbara are practised on large-scale schemes (e.g. Gezira, Guneid, New Halfa, Rahad, El Suki, and White Nile and Blue Nile Agricultural Corporation Schemes).

Water is pumped or flows by gravity from dams directly to irrigate fields where crops such as cotton, wheat, sorghum, sugar cane, groundnuts, fodder crops and vegetables are grown. Flood irrigation, using seasonal rivers such as the Gash and Baraka, Wadi Hawar and Azum, is practised in Kassala and the Red Sea and Darfur states. Water flow is directed to farmland using canals, gates and bunds (low stone or earthen walls). Crops such as millet, sorghum, cotton, groundnuts and vegetables are cultivated.

Low-rainfall savannah zone: soil and climate are diversified in this zone, and a variety of field crops are grown using different irrigation systems. Irrigated large-scale schemes, such as Gezira, extend south into this zone. Other large irrigated schemes include El Suki, El Rahad, Blue Nile Agricultural Corporation and White Nile Agricultural Corporation, in addition to large-scale sugar cane plantations in Kenana, Asalaya and West Sennar.

Traditional farming on clay soils produces sorghum and sesame, and on sandy soil millet, sesame, groundnuts, roselle and water melon are grown. This zone is also the major producer of gum Arabic from *Acacia Senegal* and *Acacia Seyal* trees. Mechanized farming is practised, particularly towards the southern part of the zone; this is mostly large-scale commercial rain-fed cultivation, where agricultural operations are partially or totally mechanised; production units are around 400 ha in size and main crops grown include sorghum, cotton and sesame. Guar has also been introduced;

Pastoralism is also a major type of land use in this zone. Pastoralists move their livestock between wet- and dry-season grazing lands (north-south-north).

High-rainfall savannah: as in the low-rainfall savannah, different farming systems such as traditional cultivation, mechanised farming and pastoralism are practised in this zone. However, the bimodal nature of the rainfall means that in the higher rainfall areas in the southwest two crops may be produced per year. Forestry is an important activity, which produces fuel wood and round wood for local industries.

Flood plain: also known as the Sudd, this is one of the largest freshwater swamps in the world, where the land is flooded to different degrees and for variable periods. The Sudd's environmental conditions make transhumance pastoralism inevitable and have also given rise to a mixed economy of herding, traditional cultivation, fishing and hunting. Main crops grown are maize, sorghum, cowpeas, tobacco and pumpkins.

Mountains: the area of Jabel Marra is used for agricultural production on a reasonable scale. It is important for field crops such as wheat and sorghum; horticultural production such as citrus, mangoes, potatoes and other vegetables; and timber (Zaroug 2006).

2-3: Water resources

Water resources in Sudan comprise three main categories (Abdalla and Karar 2010). The diagram below explains the different water resources that exist in the country.

Groundwater

Groundwater in terms of availability, quantity and quality needs further investigation and development: information on availability of groundwater

Figure 3. Water resources in Sudan

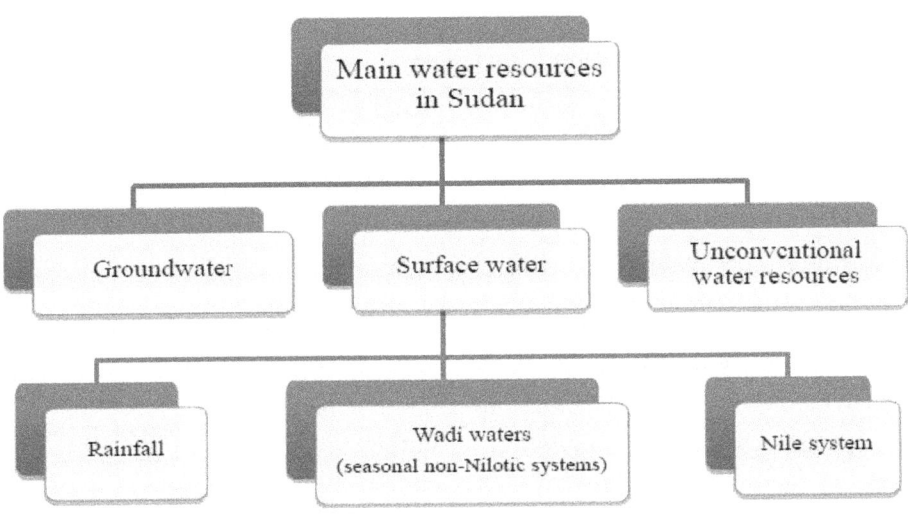

resources in the country as a whole is superficial. Groundwater resources are found in Nubian sandstones, which occupy around 28.1 per cent of country; Umm Rwaba formations and alluvial deposits occupy 20.5 per cent of the area; and basement complex formations around 42.3 per cent (Abdalla and Karar 2010).

Surface water

This is categorised by rainfall, which is generally erratic and varies according to zone; and the Nile System. Sudan shares the Nile Basin with 10 other riparian countries[3]; 63 percent of the basin falls within Sudan, and more than 70 per cent of the area of Sudan lies within the basin.

The Nile system includes the White Nile, Blue Nile and Nile rivers and their tributaries. A large part of the White Nile falls within South Sudan, which is the area of confluence of most of the tributaries of the White Nile (Salman 2011). Indeed, almost all of the important tributaries of the White Nile, including the Sobat River, either originate in or join the river there. These tributaries include Bahr el Arab (also known as the Kiir River), Bahr el Ghazal, Bahr el Zeraf, the Lol, Yei, Jur, Tonj and Naam rivers, in addition to Bahr el Jebel. The Jur and Bahr el Arab merge to form Bahr el Ghazal, after having been joined by the Lol River. Bahr el Ghazal joins Bahr el Jebel at Lake No, and thereafter the river is called the White Nile. It is subsequently joined by Bahr el Zeraf (Bitsue 2012).

---

3. Burundi, Democratic Republic of Congo, Egypt, Eritrea, Ethiopia, Kenya, Rwanda, Tanzania and Uganda.

About 28 per cent of the flow of the Nile River – which is about 23 billion cubic meters (bcm) of the total flow of the Nile out of 84bcm measured at Aswan – crosses southern Sudan into northern Sudan and eventually into Egypt. Almost 50 per cent of the waters of the White Nile are lost to evaporation and seepage in the three large areas of swampland in southern Sudan, namely the Sudd of Bahr el Jebel and Bahr el Zaraf, the Bahr el Ghazal swamps, and the Sobat/Machar swamps (Salman 2011). Bahr el Jebel itself has originates in Lake Victoria, where on exit it is called the Victoria Nile. It becomes the Albert Nile after exiting from Lake Albert, and once it crosses into it becomes Bahr el Jebel.

The Sobat River originates in Ethiopia as the Baro and Okobo rivers. It is joined by the Pibor, which originates in South Sudan, and later becomes the White Nile (Salman 2011) and flows for a considerable distance through South Sudan before entering Sudan, and subsequently merges with the Blue Nile.

The Blue Nile and its tributaries, including the Rahad and Dinder rivers, rises in the Ethiopian highlands. The White Nile and the Blue Nile merge at Khartoum to form the Nile River (WHO 2009). The Nile is joined in the north by the Atbara River, which also originates in the Ethiopian highlands. The Atbara is the last tributary to join the Nile, which thereafter flows through northern Sudan and Egypt before it enters the Mediterranean Sea.

The Ethiopian plateaus are the origin of about 86 per cent of the waters of the Nile; the Equatorial Lakes contribute a further 14 per cent of the total flow (Lean 2009). Despite the high contribution of the Blue Nile, its peak flow is largely seasonal, concentrated mostly in the months of June-September. The relatively smaller contribution of the White Nile, however, is mostly steady throughout the year, which provides the critical water needs of Sudan and Egypt during the low flow period of the Blue Nile.

Thus, the two rivers complement each other and provide for a perennial river in Sudan and Egypt. It should also be added that the Blue Nile is quite heavy with silt that it carries from the Ethiopian highlands, whereas the White Nile is almost silt free (Salman 2010). The hydrology of the Nile and its tributaries is summarised in Table 6 below.

Wadi waters

Among the large number of seasonal rivers and water courses, Gash (Mareb) and Baraka are the largest wadis, each with an annual average flow ranging from 200mcm to 800mcm, which occurs between July and September. These wadis originate in Eritrea and terminate in the continental deltas of Gash, Toker and Red Sea, and are important for flood irrigation agriculture (Ahmed and Ismail 2008).

The Azum, Hawar, Kaja, Ebra, Toal, Elkou and Salih are the largest wadis in western Sudan, with estimated annual run-offs ranging from 120mcm to

Table 6 Water supply from the Nile and its tributaries

Tributary	Total annual average supply (BCM)	Flow characteristics
Blue Nile	50.7	Average daily peak discharge falls from 535mcm/day in August to 11mcm/day in April
Rahad	1.09	Flow from July to November
Dinder	3.0	Flow from June to November
White Nile	27.8 (at Malakal)	Daily flow falls from 114mcm/day in November to 54mcm/day in April
Bahr el Gazal	14	0.5bcm reaches Malakal (swamps)
Bahr el Jebel	26 (at Mongalla)	Flow falls from 66mcm/day in November to 8mcm/day in April
Sobat	13.3 (at Malakal)	Losses in Baro and Machar reach 8bcm. Flow falls from 66mcm/day in November to 8mcm/day in April
Atbara	12 (7 from Setit and 5 from Atbara branch)	Low regulated flows from February to June
Main Nile	84 (at Aswan)	Average daily peak flow of 690mcm/day (August-Sept.) and a low flow of 74mcm/day (April-May)

Source: (Abdalla and K. 2010)

500mcm (Abdalla and Karar 2010). The combined total annual run-off of all the wadis is estimated to vary between 5bcm and 7bcm (Abdalla and K. 2010). However, the total amount of fresh water from internal and external sources is around $30 \times 10^9 m^3$/year, which brings per capita water availability below the water stress limit of $1000m^3$ (FAO/RNE 1998). The table below summarises the water available to Sudan.

To use the available water, Sudan has constructed five dams that are mainly used for generating hydropower and supplying irrigation water to four main gravity-fed irrigation schemes: Gezira (established in 1925), New Halfa (1964), Suki (1971) and Rahad (1977) (Ahmed and Ribbe 2011). Evaporation is generally high, ranging from 1000mm/year to 3000 mm/year. This affects the country's arable land and reduces its capacity to produce crops for domestic consumption and export. The irrigated area would be less if some of it was planted to perennial

Table 7 Summary of the available water to Sudan

Water resources	Quantity (bcm)	Constraints
Sudan's present share from the Nile Waters Agreement (at central Sudan)	18.5	Seasonal patterns coupled with limited storage vessels; expected to be shared with riparian neighbours.
Wadi waters	5–7	Highly variable, short-duration flows, difficult to monitor or harvest; some shared with neighbours.
Renewable groundwater	4.0	Deep water entails high cost of pumping; weak infrastructure in remote areas.
Present Total	**30.0**	
Expected from reclamation of swamps	6.0	Capital investment needed; considerable social and environmental costs.
Total	**35.5–37**	

Source: Eltom et al. 2000

Table 8 Main storage reservoir capacities (km³)

Dam	Design capacity	Actual capacity	Established
Sennar	0.9	0.4	1925
Roseires	3.4	1.9	1966
Khashm el Girba	1.3	0.5	1964
Geblawlia	3.0	3.0	1937

Source: (Ahmed and Ribbe 2011)

crops or used for more than one seasonal crop per year. In addition to irrigation, water is needed for municipal and industrial uses, which are increasing with the expansion in urbanisation and industrialisation. The water stress situation is therefore bound to become even worse (FAO/RNE 1998).

Rainfall in Sudan decreases from south to north, the annual average varying from 120 cm (47 in) in the south to less than 10 cm (4 in) in the north (Mohamed-Ali, Luster-Teasley et al. 2009). The country suffers from many ecological crises such as a chronic shortage of freshwater overall the country (UNEP 2007), drought, and desertification (Barton and Writer 2012). Furthermore, water is not equally distributed all over the country; there are major regional, seasonal and annual variations. Underlying this variability is a creeping trend towards generally drier conditions (UNEP 2007). Moreover, between the mid-1970s and late 2000s, summer rainfall decreased by 15%–20% across parts of the west and south. These declines can be visualised as a contraction of the region that receives adequate rainfall for viable agricultural livelihoods (Elgali, Mustafa et al. 2010).

Rural Sudanese are forced off their land by changing landscapes and a lack of agricultural production. Demand for water is increasing, but its availability to the country's inhabitants remains low (Barton and Writer 2012). About 2 percent of water is available for domestic use; in comparison, in the US, water for domestic use accounts for 13% of total supply (The Water Project 2012). Data from the Sudan Household Health Survey 2006 show that about 40 per cent of the population does not have access to safe drinking water and more than two-thirds have no access to adequate sanitation (UNICEF 2010). Access to water is critical, because many regions have become neglected or are experiencing conflict, such as Darfur and other regions.

Water availability and accessibility varies by region. The Butana region is on a dry plateau east of the Nile and has a reputation for prime grazing land; but rainfall in the area is limited to a brief period in the middle of the year and the vegetation is highly seasonal. The area is crossed by many transhumance routes. Hence, signs of rangeland degradation are most apparent in areas around sources of drinking water, which are extremely localised. Streams and ponds are quickly drained after the rains because there are no perennial rivers. Livestock

Image 1. Water pump, Waw, South of Sudan

and people must rely on a variety of specialised man-made water sources. The most widely used are traditional dugout reservoirs (*hafirs*), which harvest water from surrounding land during the rainy season and are the most common man-made source of dry season water (Cox 2010).

In semi-arid zones such as the west, safe drinking water is rare. In particular, areas in North Darfur and South Kordofan states rely on groundwater supplies (wells) or hafirs as in Butana. Another traditional method of saving water that is used in this zone is to keep water in baobab trees (*Adansonia digitata*, locally called *Tebaldi*) to use in periods of drought. Farmers in semi-arid areas in the east save water for agriculture by building *teras* plots (Practical Action 2012).

The *teras* is "bunded" on three sides and the fourth side is left open to capture runoff from an adjacent, slightly elevated catchment. The bunds consist of small stone or earthen walls, constructed along the plot contour to obstruct overland water flow on hill slopes. These bunds reduce flow velocity, so that water percolates into the soil behind them, increasing soil moisture and recharging groundwater. Teras show higher crop returns in drier years and allow farmers to diversify income sources in normal years. In West Africa the technology is widely used in valley bottoms (Netherlands Water Partnership 2007).

According to FAO, agriculture is the single largest user of freshwater on a global basis and a major cause of degradation of surface and groundwater resources through erosion and chemical run-off (Ongley 1996). Agriculture

Table 9: Sudan: Total water requirements, 1957

Region	Feddans[1]	m³ per feddan	Billion cubic meters
Southern Sudan	500,000	2700	1.35
White Nile pumps: Sobat-Geiger	120,000	3400	0.41
White Nile pumps: Geiger-Khartoum	210,000	4400	0.92
Blue Nile pumps: Roseires-Sennar	250,000	3600	0.90
Blue Nile pumps: Sennar-Khartoum	200,000	4500	0.90
Kennana gravity area: west	600,000	4400	2.64
Kennana gravity area: east	580,000	3600	2.09
Kennana gravity area: south	40,000	3100	0.12
Gezira and Managil	1,800,000	4500	8.10
Butana	500,000	4500	2.25
Northern Sudan pumps	700,000	5330	3.73
Total	5,500,000		23.41
Less 10% for transmissions losses			2.34
As at Aswan			21.07

Source: Taha 2010

in Sudan accounts for 97% of the country's water use (Sullivan 2010, Barton and Writer 2012). The diversion of water to mechanised farms and intensive cultivation by rural farmers is contributing to the spread of arid conditions cross Sudan (Barton and Writer 2012). Water is in high demand to meet the needs of rapid population growth and food production, and plans to expand agriculture through irrigation further raises the demand for water. Table 9 shows Sudan's estimated water requirements in 1957 (Taha 2010).

Despite the oldness of Table 9 it is included in this report to serve two purposes: First, to explain that the Sudan's requirements of water at that time- according to the table- exceed its share of the Nile waters by 3bcm. That means 'The 1959 Nile Water Agreement' between Egypt and Sudan was not fair. And the water amount allocated to Sudan would not be enough if the country would have implemented all entries in the table; Second, Sudan would face water deficit if it implemented its policy to extensively increase the area of irrigated land. tried to expand in its water and agricultural projects, bearing in mind drought spells that Sudan faces recently from time to time.

The government has estimated that the country's water requirements for food security and other essentials will be 32bcm by 2025. However, at present Sudan is not using its allowed share (Taha 2010).

According to Salman, a water law and policy expert, Sudan is using only 12bcm out of its share of 18.5bcm. Thus, Sudan has failed to use 6.5bcm annually for 54 years since 1959. That amounts to 350bcm that Egypt has used (Salman 2013). About 80 per cent of irrigated schemes in Sudan were developed in the

---

1. 1 feddan = 0.42 ha

early 1960s. They were designed on the basis of a constant water calculation now considered defunct: 400m3/feddan applied at 14-day intervals. Any unused amount of water by Sudan is used by Egypt as loan.

Referring to ANNEX 1 in Nile Water Agreement, there is "A Special Provision for The Water Loan Required by The United Arab Republic" which states the following

The Republic of the Sudan agrees in principle to give a water loan from the Sudan's share in the Sudd el Aali waters, to the United Arab Republic, in order to enable the latter to proceed with her planned programmes for Agricultural Expansion.

The request of the United Arab Republic for this loan shall be made after it revises its programmes within five years from the date of the signing of this agreement. And if the revision by United Arab Republic reveals her need for this loan, the Republic of the Sudan shall give it out of its own share a loan not exceeding one and a half Milliards, provided that the utilization of this loan shall cease in November, 1977.

However, the loan has not been ceased until today. That Annex will support Sudan's position when bargaining to get this loan repaid in another form (Taha 2010). Water requirements will become severe if we consider environmental factors, such as increased desertification and land degradation, which have intensified Sudan's water problems (Ashok 2008).

Part 3: Agricultural schemes in Sudan

3-1: Major agricultural schemes

Sudan has the largest irrigated area in sub-Saharan Africa and ranks second only to Egypt on the continent in terms of irrigated agriculture. Commercial agricultural activities are mostly concentrated in a belt across the centre of the country, known as the central clay plain, which extends approximately 1100km from east to west between latitudes 10° and 14° north, in the arid and semi-arid dry savannah zone (UNEP 2007).

Map 4. Agricultural Schemes in Sudan

Irrigated Agricultural Schemes
1. Gezira and Managil — 870'750 ha
2. New Halfa — 152'280 ha
3. Rahad — 121'500 ha
4. Gash Delta — 101'250 ha
5. Suki — 35'235 ha
6. Tokar Delta — 30'780 ha
7. Guneid Sugar — 15'795 ha
8. Assalaya Sugar — 14'175 ha
9. Sennar Sugar — 12'960 ha
10. Khashm El-Girba — 18'225 ha
11. Kenana Sugar — 45'000 ha

Mechanized Agricultural Schemes
(planned and unplanned)
1. Habila
2. El-Dali
3. El-Mazmum
4. El-Raheed
5. El-Sharkia
6. Dinder
7. Gedaref
8. Southern Kordofan
9. White Nile
10. Upper Nile
11. Blue Nile

Agricultural schemes boundaries are approximate.

The boundaries and names shown and the designations used on this map do not imply official endorsement or acceptance by the United Nations

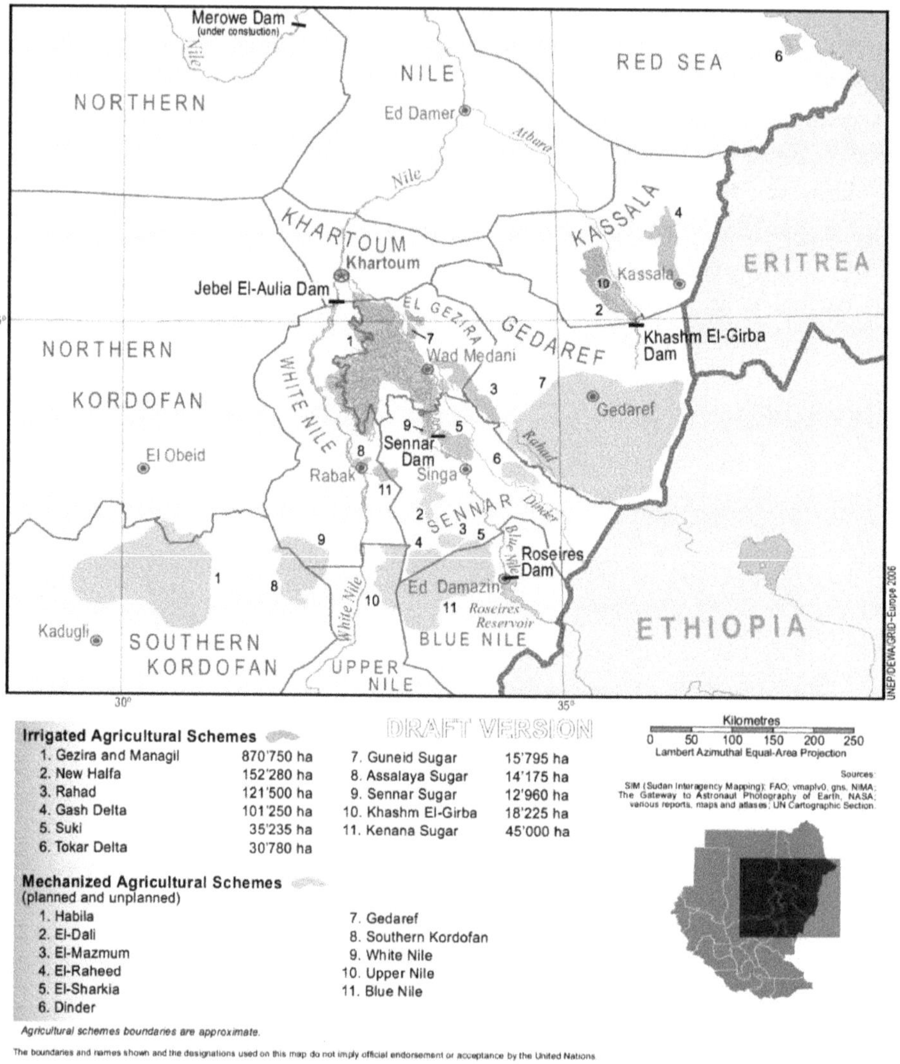

Source: (UNEP 2007)

Agriculture is divided into two main sectors: irrigated and rain fed. The irrigated sector covers about 1.8 million ha and includes the Gezira, Rahad, New Halfa, Elssuki, White Nile and Blue Nile schemes. Gezira, Rahad and New Halfa are considered the most important, and they produce cotton, groundnuts, wheat, sorghum and vegetables (Mahir and Abdelaziz 2010).

3-2: Irrigation schemes

The irrigated sector in the Sudan broadly falls into traditional and modern. Approximately 90 percent of the irrigated area is modern (UNEP 2007). Irrigated agriculture comprises three main categories – gravity, pump and basin (flush) – as well as some small basins (Ahmed, Sulaiman et al. 2012). Although irrigation only covers 7 per cent of the cultivated area, it accounts for more than half of crop yields (UNEP 2007).

Large-scale irrigation schemes have been Sudan's leading economic investment, but various studies indicate that their performance has been considerably below potential. For example, of the 1.9 million ha prepared for irrigation in 2005, only half was actually cultivated, largely because of decrepit irrigation and drainage infrastructure (UNEP 2007); low producer prices; lack of foreign currency; and import regulations, which have limited the availability of vital production inputs and spare parts (IFAD 1992). Environmental factors such as canal sedimentation have also contributed to low irrigation returns (UNEP 2007). The Gezira and New Halfa Schemes are gravity irrigation schemes.

3-3: The Gezira Scheme

"The opening page of Gaitskell's book is a map. This is not an ordinary map, however, but one that shows key features of the Gezira region set within the map of England. On this map, the Blue Nile runs from the Sennar Dam at London to meet the White Nile, some distance east of Liverpool, at Khartoum. While the ostensible function of the map is to acquaint British readers with the vast scale of the Gezira, this is not all it accomplishes. By relocating the Gezira in England, the map simultaneously claims the Gezira for Britain and removes it from any larger Sudanese context (Bernal 1997).

The Gezira Irrigation System, British Sudan, 1900–49

Two main reasons were behind the British occupation of Sudan in 1898: to control water resources to ensure water supplies from the Nile for cotton farms in Egypt, which was of utmost importance for the British textile industry; and to safeguard the Suez Canal and the route to India by controlling the region from Cairo to Cape Town (Ertsen 2006).

In 1904, Sir William Garstin, the under-secretary of state for public works in Egypt, published the first overall plan for control of the waters of the entire Nile Basin. The plan included, among other things, increasing the storage capacity of the Aswan Dam; using the southeastern Gezira region as a wheat-producing area for the nearby Arabian market, with only a small emphasis on cotton; and the building of a dam or barrage at Sennar on the Blue Nile to provide irrigation for part of Gezira (Ertsen 2006).

In 1900, the Governor-General of Sudan Sir Reginald Wingate proposed starting irrigated cotton production in Gezira and using irrigation revenues as a source of income to finance the increasing expenditures of his government. The Gash and Tokar deltas in eastern Sudan were test areas for growing cotton. In 1904, US businessman Leigh Hunt founded the Sudan Experimental Plantations Syndicate (SEPS), which was granted a cotton concession for 10,000 feddans [4] at Zeidab, 180 miles north of Khartoum along the Nile (Gaitskell 1959). The results at Tokar, and especially at Zeidab, showed that quality cotton production was possible in Sudan. In 1911, the Tayiba pump-irrigation project was established to grow long-staple cotton and dura sorghum (Ertsen 2006).

The vast irrigation project drew the attention and endorsement of the British government. The British cotton industry lobbied the government to guarantee a loan of £3 million in 1913. The outbreak of World War I in 1914 interrupted plans (Omer 2011) and costs increased sharply. In 1919, the area to be irrigated was set at 300,000 feddans. A delegation from the British Cotton Growing Association visited Gezira that year to promote the construction of the Sennar Dam and the irrigation system. A total of £13 million was finally reserved for the project: £11.5 million for the dam and 300,000 feddans of land, £700,000 for railways and £400,000 for cotton ginneries (Gaitskell 1959).

In 1922, the contracts for the construction of the dam and the main canals were awarded and the work started. Already confronted with huge cost overruns, but fearing the political and economic effects of abandoning the project, the British government guaranteed loans totalling almost £15 million in 1924. The following year, the High Commissioner for Egypt and Sudan Lord Lloyd, officially opened the dam (Ertsen 2006).

In the meantime, the government and the SEPS had defined the terms under which the Gezira Scheme would function. In 1919, a tripartite partnership – with the 'tenants' as the third party – was formulated, which laid out the partner's responsibilities and benefits (Ertsen 2006). The Sudan government was responsible for Sennar Dam and major work on the irrigation canals and would receive 40 per cent of total net profits from cotton production. The government rented the land from its official owners and rent it annually to tenants.

---

4. 1 feddan = 0.42 ha

Table 10. Irrigated area of the Gezira Scheme

Year	Irrigated area (feddans; 1 feddan = 0.42 ha)
1926	300,000
1929	379,000
1931	527,000
1953	1,000,000

These tenants, organised in farming units of 20–30 feddans, received 40 per cent of net profits, but had to finance the production costs through loans until the cotton was delivered (Salman 2013). As the concession holder, the SEPS was responsible for canal-cleaning and land-levelling, management of the scheme, and providing field and accountancy staff, buildings, loans to farmers, and transport of the cotton; in return, it received 20 percent of net profits (Gaitskell 1959).

English thought to expand the established 300,000 feddans so, more land was needed. The Gezira Scheme had already alarmed Egypt, which feared two things. Firstly, the amount of water used to irrigate the scheme jeopardised the amount of water that would reach Egypt. Secondly, the competitiveness in international cotton market from Sudanese cotton produced in the Gezira Scheme was bound to increase (Salman 2013).

Following negotiations, the British government committed to guarantee Egypt's water supply and to pay compensation for water used to irrigate Gezira. That cleared the way for the Nile Waters Agreement of 1929, which allotted 48bcm to Egypt. In addition, during the dry season, months from December up to July, the whole flow of the Nile was reserved for Egypt and after 15 July Sudan was allowed to water the Gezira Scheme and fill up the Sennar Dam (Tvedt 2004).

Subsequent extensions steadily increased the area under irrigation, as the table above shows:

In 1944, it became clear that the SEPS concession to operate the Gezira Scheme would not be renewed in 1950. In July 1949, a government board assumed management of Gezira (Ertsen 2006), with some 80,000 tenant households. Besides administration, the board provided credit, marketing and other services (Salman 2013). One of the main objectives of the scheme was social development; hence, an small portion of the profit (2%) was directed to finance social development projects throughout Gezira province (Hassan 2011).

The economic and social development successes of the Gezira Scheme in early 1950s were behind plans to expand its area. But it was also the beginning of conflicts of interest and competition over the Nile and economic power between Sudan and Egypt (Salman 2013). Disputes continued for more than half a century from the start of Anglo-Egyptian rule on Sudan up to the signing of the Nile Water Agreement in 1959.

The agreement governed the full use of Nile waters (84bcm annual on average as measured at Aswan); Egypt's share of water increased to 55.5bcm and Sudan's share was 18.5bcm, the rest being lost to evaporation in Lake Nubia. The agreement cleared the way for irrigating the Managil Extension, which extended the original Gezira Scheme over 1959–63. The combined Gezira/Managil Scheme now covers an area of about 2.3 million feddans (Salman 2013)[5].

The Gezira Scheme is the oldest and the largest – in terms of area – irrigated scheme in the history of Sudan, and its most important development project. It is also used to be the biggest irrigation system under one administration in the world (Bernal 1997). It is located in Al Gezira state (which means island or peninsula) southeast of the confluence of the Blue and White Niles at Khartoum.

Certain features made the area on which the scheme was established the most suitable location. First, the area is situated between two rivers, which facilitated the construction of dams at Sennar and Roseires that provide abundant water for irrigation. Second, the is geographically and topographically suited to irrigation: the land slopes westwards away from the Blue Nile and northwards away from the Sennar Dam; water therefore naturally runs through the irrigation canals by gravity[6], which facilitates cheap irrigation rather than costly artificial methods (Salman 2013).

Also, the soil has a high clay content, which is fertile and keeps down losses from seepage (Al-Naiem 2009, Omer 2011). High summer temperatures are another advantage, because they reduce insects and other agricultural pests, hence there is no need to use expensive pesticides. Furthermore, the area is situated in the centre of the country and benefits from seasonal labour movements to and from regions in Sudan and neighbouring countries and proximity to the harbour at Port Sudan (Salman 2013).

Until its collapse, the government appointed the scheme's board of governors, which the minister of agriculture and forestry chaired. A general manager was at the head a staff of around 7000 employees in 2000 (World Bank 2000), which was later considerably reduced to 400 (Salman 2013).

Until the 1990s, the scheme played an important role in the country's economic development, and was a major source of foreign exchange earnings and government revenue. It occupied a central position in the agricultural sector and produced the bulk of the country's cash crops (Al Naiem 2009). Moreover, it also contributed to national food security and was the basis of the livelihoods of some 2.7 million people who lived within the scheme (Al Naiem 2009).

---

5. In 2000, the scheme accounted for about half the area of irrigation schemes that drew water from the Nile system (World Bank 2000). In 2004, it used about 8bcm of Nile water, 35 per cent of Sudan's allocation (Eldaw 2004).
6. Gravity irrigation takes advantage of the natural slope of the land (gradient) without pumping and/or lifting.

The organisational structure of the Gezira Scheme is divided into 18 'groups' and some 100,000 tenancies, each holding (about 8ha). The 18 groups range in size from 60,000 to 190,000 feddans. Each group consists of smaller 'blocks', which comprise 'numbers', each of 90 feddans (Al-Naiem 2009). The socio-economic composition of the scheme is about 128,000 tenant households, which constitute with their families about 1 million people. In addition, there are some 150,000 seasonal labourers (who with their families constitute more than 1 million people); and 10,000 employees and permanent workers (until their recent, reduction, in 2005, to about 400) (Salman 2013).

About 55 per cent of the land is government owned; the remainder is owned by landholders with whom the central government has a long-term rental agreement. There have been some major disputes between the owners and the government over rent, and cases are pending before the courts (Salman 2013). The scheme's assets until recently included service centres, which comprised: 14 scutchers; 18 engineering workshops; 1300km of railway; a telecommunications network; a fleet of tractors, combine harvesters and other vehicles; 444 stores with a total capacity of 25 million metric tons; more than 6000 residential units and 76 compounds; 200 offices; 78 schools and health centres; 53 water purification facilities; a road network; a research centre; and 150,680-km irrigation network.

As mentioned above, management of the scheme was shared between the government, the tenants and the board. From the start of the scheme in 1925 until 1950 profits were distributed according to 40:40:20 ratio. When the Sudan Gezira Board took over management in 1950, the ratio changed to 42:42:10, with the rest for public services. In 1965, as reward to the tenants for their role in the revolution that successfully ousted General Ibrahim Abboud's military junta from power, their share increased to 48%; the government's share was reduced to 36% (Salman 2013).

Before mid-1970s, the scheme followed a system of farming whereby only cotton received official attention; sorghum and beans were grown as farmers' crops and necessary components of the crop rotation cycle. After then, a diversified farming system was adopted, and in addition to cotton farmers were encouraged to produce wheat, sorghum, groundnut and vegetables in an effort to make Sudan self-sufficient in foodstuffs (Omer, 2011).

Crop rotation was changed several times in accordance with the prevailing agricultural environment that coincided with the different stages of development in the scheme (Hassan 2011). Initially, the tenants practised a six-course rotation. Each tenant had to plant according to the approved rotation so that, for example, all the cotton grew at the same time. Farmers practised a four course-rotation (cotton, sorghum, legume, fallow) in the 1950s. Sorghum, the staple food of the tenants, though exhaustive to the soil nitrate-nitrogen, was included in the rotation to give security of tenure to the tenants (Hassan 2011).

In the early 1980s, they adopted an eight-course rotation (cotton, fallow, fallow, cotton, fallow, sorghum, cowpea, fallow), with a nominal cropping intensity of 50 per cent. This kept the demand for water within the capacity of the irrigation system. Since then, there has been further diversification and intensification (Al-Naiem 2009). Five-course rotation was adopted all over the scheme with the aim of achieving the best results from the available resources of land and water (Hassan 2011). However, this crop rotation was radically changed by the freedom of crop choice under the Gezira Scheme Act 2005 (see below).

After more than 70 years, in 1998 the government judged the Gezira Scheme to be inefficient and a drain. Cotton production in the Gezira Scheme had declined and the value of cotton exports, which used to provide about half of total export earnings, were second to livestock (World Bank 2000). According to Salman (Salman 2013), the scheme began to collapse in the 1970s for the following reasons:

- Poor maintenance of the irrigation canal network led to serious water management and distribution problems; and the use of river water with a large sediment load and extensive weed growth clogged up the canals, all of which reduced irrigation efficiency. Gradually, water shortages or drought became the main feature of the scheme instead of cotton.
- The situation was worsened because the reservoirs of Sennar and Roseires dams lost half of their storage capacities as a result of siltation.
- The scheme became trapped in a vicious circle: the lack of water for irrigation because of siltation in the irrigation led to deterioration in crop production, hence, high production costs and low profits, and so on.
- During the 1973 oil crisis, the government failed to provide the required agricultural inputs, which resulted in low yields.
- Low cotton prices because of increased supply on the international market from Asia, China and India.
- Costly pest control for cotton; inadequate financing and marketing arrangements for most crops; inefficient agricultural processing; disillusioned farmers and low cost recovery for irrigation water deliveries.
- The scheme regularly incurred an operating loss and tenants often found it impossible to make ends meet from farm income alone, despite the fact that the average farm size was 20 feddans (Elageed 2008).
- Inconsistent agricultural policies.
- Failure of all rehabilitation attempts by different donor agencies because of a lack of institutional reform in the economic, social and managerial structures of the scheme.

A presidential committee examined the operation of the scheme and recommended, among other things, that the scheme should be converted into a joint stock company, in which the government and the private sector would have shares. Intense national debate followed (World Bank 2000). In June 2005, the National Council passed the controversial "Gezira Scheme Act of 2005"[7], which the president assented to the following month. The Act introduced many changes:

- The old financing system, under the control of the Central Bank and the Ministry of Finance, was abolished and substituted with a new one based on the international free market and privatisation. (Hassan 2011).
- The government assumed control of scheme assets, with a view to future private sector participation, and investment either in existing assets or new additions to the scheme.
- The composition of the scheme based on shared participation between tenant households, the SGB and the government has changed. It would henceforth be composed of:

A. Farmers.

B. The government, which would provide basic services such as development, irrigation and public goods, including: research, plant protection, technology support, agricultural extension, technical studies, and training, as well as supervisory management and indicative planning.

C. The private sector, which would provide auxiliary commercial services.

- The production relationship system used to be based on land and water rates for each crop to tenants individually, with the SGB in its role as landlord operating and maintaining the lower reaches of the irrigation system. The Act gave sweeping new powers to water users associations (WUAs) and the private sector, and significantly reduced the role of the public sector (Mohamed Alameen 2009). Under the Act the changes are as follows:

    a. The Ministry of Irrigation and Water Resources shall be responsible for the operation and management of the primary irrigation and drainage canals and pumps in the scheme, and for providing sufficient water for WUAs at the mouths of the respective field canals; and the Ministry of Finance and National Economy shall be responsible for financing the maintenance, rehabilitation and operations of water canals in return for water charges to ensure provision of such services.

    b. WUAs shall maintain, operate and manage field canals and internal drainage.

    c. All irrigation operations for any part within the scheme command area shall have to be approved by the Board.

---

7. The Act is included in the appendix. The Act was promulgated in Arabic; the English translation was provided by Dr Salman M. A. Salman.

In 2010, the Sudanese media reported that, in a dramatic shift in government policy, the Ministry of Agriculture had signed an agreement with the Egyptian Ministry of Agriculture to develop 1 million feddans of the scheme – half of the total area – in accordance with a provision of the Sudanese Egyptian Protocol of Cooperation (Ali 2010). Public and private Egyptian corporations would provide all the necessary inputs and the agricultural produce would then be exported to Egypt according to a very strict schedule. Any delays would be penalised.

This led to accusations of land grabbing. The Sudanese Farmer's Union, the Gezira Scheme tenants and commentators in the Sudanese media refused to accept the deal (Ali 2010). They argued that no activity could take place before the government had settled issue of the land ownership in the scheme. Some described the deal as a betrayal of the tenants, farmers and the Sudanese people, given the significant role that the scheme had played over the years. Under this pressure, the agreement was eventually abandoned (Ali 2010).

A committee was established in February 2013 to review the situation in the Gezira Scheme and make recommendations on how the scheme should be organised and operated. The committee presented its report in May, but the report has not been made public and the government has not implemented its recommendations (Salman 2013).

3-4: New Halfa Irrigation Scheme

Background

Sudan made large and successive concessions to Egypt during five years of negotiation on the Nile River waters, which lasted from 1954 until the signing of the above-mentioned convention on Nile waters on 8 November 1959. The concessions included approval of the submerging town of Wadi Halfa and 27 village communities between Aswan and the Dal Cataract in northern Sudan. Also submerging more than 200,000 feddans of fertile agricultural land, more than 1 million palm and citrus trees, huge amounts of minerals and precious metals (iron and gold), as well as the Dal and Semna waterfalls, with the potential generate more than 650 MW of electricity[8].

---

8. Article 6 and 7 in Nile River agreement signed between Sudan and Egypt in 1959 read as follow:
    6. The United Arab Republic agrees to pay to the Sudan Republic 15 Million Egyptian Pounds as full compensation for the damage resulting to the Sudanese existing properties as a result of the storage in the Sudd el Aali Reservoir up to a reduced level of 182 meters (survey datum). The payment of this compensation shall be affected in accordance with the annexed agreement between the two parties.
    7. The Republic of the Sudan undertakes to arrange before July 1963, the final transfer of the population of Halfa and all other Sudanese inhabitants whose lands shall be submerged by the stored water.
    http://www.internationalwaterlaw.org/documents/regionaldocs/uar_sudan.html

Egypt offered to pay compensation to affected Nubian families (Salman 2013), which is detailed in point 6 of the agreement: "The United Arab Republic agrees to pay to the Sudan Republic 15 Million Egyptian Pounds as full compensation for the damage resulting to the Sudanese existing properties as a result of the storage in the Sudd el Aali Reservoir up to a reduced level of 182 meters (survey datum). The payment of this compensation shall be affected [sic] in accordance with the annexed agreement between the two parties" (Sudan and Egypt 1959, p.2). However, the whole total cost of the forced deportation process exceeded 37 million pounds (Salman 2013).

Around 120,000 indigenous Nubians (70,000 in Egypt and 50,000 in Sudan) were displaced (Salman 2013). The Egyptian Nubians were forcibly resettled on newly reclaimed lands near Kom Ombo, 45km north of the city of Aswan. The Sudanese Nubians, including 11,000 inhabitants of Wadi Halfa town were forcibly relocated to Khashm el Girba area, some 850km southeast of their original homes (Sørbø 1985). They were settled on agricultural lands developed as a result of the Aswan Dam agreement.

The construction of a dam on the Atbara river began at Khashm el Girba village in 1961, and the majority of the Nubians had been moved to New Halfa by 1964. Later, a large number of nomadic and semi-nomadic inhabitants of the area were also established as resident farmers on what came to be the second-largest irrigation project in Sudan, the New Halfa Agricultural Production Scheme (Sørbø 1985).

The New Halfa Agricultural Production Scheme

The New Halfa area is located at 15° 21' N and 35° 37' E in Kassala Province, 500km from Khartoum. It was established in 1964 on a large plain in the Butana area, to the west of the Atbara river and north of the village of Khashm el Girba. It stretches about 100km in a north-northwest direction and is between 20km and 35km wide. The region is within Sudan's semi-arid dry savannah belt, with annual rainfall ranges between 250mm and 300mm (Sørbø 1985, Himeidan, Hamid et al. 2007).

The Gezira Scheme was the model for New Halfa Agricultural Scheme (Davies 1985). It was designed to use water from the Khashm el Girba reservoir on the seasonal Atbara River. The initial reservoir capacity was 1.3bcm. The average annual inflow was about 12,000 million m^3 varying between nil in March-May and a peak of 5-7bcm/month in August. The reservoir capacity was rapidly reduced by siltation in just 12 years to about 0.8bcm in 1976 (World Bank 1978).

Despite flushing measures that were introduced in 1971, the reservoir storage capacity continued to decline and was projected to fall by 6 per cent per year. In an attempt to reduce the loss, the Ministry of Irrigation has increased the

Table 11. Composition of New Halfa Production Scheme by land use

Land use	Area (feddans)
22,000 tenancies	330,000
Halfawyeen freehold land	24,000
Reserved areas	19,000
Research sub-station	1000
Afforestation	2500
Sugar estate	41,000
Infrastructure and waste	29,500
Total	**447,000**

Source: World Bank (1978)

maximum normal storage level by 1m to 474m above sea level. The scheme relied on gravity flow to distribute water to the project area from the dam during peak times; otherwise, from about May to August the reservoir level falls below the canal level and water had to be pumped through a canal system (World Bank 1978). However, water losses were great and irrigation efficiency low, which decreasing dam capacity because of siltation exacerbated (Davies 1985).

The scheme area comprised 447,000 feddans of almost flat clay or loamy soils, of which 330,000 feddans each year were under three main crops – cotton, wheat, and groundnuts – grown in an annual rotation with a maximum area of 110,000 feddans each. The average cropping intensity was in the range of 91 per cent for cotton, 88 per cent for wheat, and 48 per cent for groundnuts. Dura gradually replaced wheat or groundnuts among the nomad tenants from 1979. Cotton, wheat, groundnuts and sorghum were grown on the tenancies, and vegetables and fruit on the freehold land (Davies 1985). The project area was divided as follow (World Bank 1978):

After completion of the resettlement of the 7000 Halfawyeen families, the remaining tenancies were allocated to local pastoralists (nomads), making a total of 22,000 tenants. In 1978, the population in the project area was heterogeneous and estimated at about 300,000 people, of whom 68,000 were Halfawyeen, 148,000 from different nomad tribes, 50,000 migrant labourers and 34,000 inhabitants of New Halfa town.

The Nubian population was distributed over about 25 villages (Sørbø 1985) and the nomads concentrated in 51 villages (38 inside the scheme and 13 on its fringes). These villages were poorly equipped in terms of essential services. It was very obvious that there was little integration, if any, between the Halfawyeen and the nomads and other ethnic groups in the area. They had different origins and customs and, because of this, each community was spatially segregated in special villages of its own and settled separately. The staple food crop of Halfawyeen is wheat and of the others sorghum; this explains why the dura replaced wheat in nomads tenancies, as mentioned above (World Bank 1978, Davies 1985).

As with the Gezira Scheme, management was a partnership between the government, the New Halfa Agricultural Production Corporation (NHAPC) and the tenants. NHAPC is responsible for supplying goods and services to the tenants, and has a board that represents the various parties. The tenants have their own organisation in the form of a tenants union. They cultivate their 15-feddan farms in accordance with prescribed cropping patterns, and use family and hired labour. They repay the corporation for the goods and services it supplies out of crop earnings (World Bank 1978).

Objectives of the scheme
The project was designed so that its main objectives reflected national policy. These are summarised below (Davies 1985):
1. To resettle 52,000 Nubians and compensate them for their land submerged by the reservoir behind Egypt's Aswan High Dam, following the 1959 Nile Waters Agreement between Sudan and Egypt:
2. Sedentarisation of Butana nomads, by providing them with tenancies, because they lost their grazing lands to the scheme.
3. To use the Butana plains and Atbara river waters for intensive cultivation and expansion of modern agriculture, and the future development of agro-industry.
4. To increase production of cotton and groundnuts for export, and of sugar and wheat to avoid importing such strategic food crops, hence contributing to the country's balance of payments.

Human dimensions as a key factor in the failure of the project
One could argue that the New Halfa Agricultural Scheme carried the seeds of its own failure from its inception. What follows are some accounts of the Nubian exodus and the nomads' land transformation. These images reveal vividly that the human dimension was ignored in the process of displacement and not taken into consideration when the project was planned. This turned out to be one of the most important reasons for its failure. (Ali 2012).

Ali from Khartoum University described it thus:

> "The weak economic performance of the new Halfa scheme, and the failure of the scheme to attain its prescribed goals, was mainly caused by the inherent weakness of the package of "resettlement" as a development strategy, especially in the long-run. Uprooting populations from their original environments to compel them to live in strange natural and socio-cultural settings often leads to undesirable ends; hence the social cost of such enforcement is certainly, too high, and can never be compensated. Ethical issues add to the complexity

of resettlement, because of the human rights involved in resettlement and particularly in light of the fact that it is often of a forced nature and often gives rise to unexpected outcomes. Therefore, development schemes which necessitate construction of dams and relocation should be revised; and as local experiences proved failure of sedentarization of nomads long time ago, resettlement experiments also seem to face the same fate, not only because of the infeasibility of dams as long-sustaining development solutions, but also due to the fact that the issue contains some aspects of rejecting others' right to choose what is good for them, and to avoid misuse of valuable resources." (Ali 2012 p.1)

This was in line with Sørbø (Sørbø 1985), who portrayed a situation that reflected the importance of the human dimension that had been neglected, both in the displacement of Nubians from their ancestral land; and in the planning of the new resettlement area for them:

"There were immense problems of adjustment for the Nubians as well as for the nomads. The Nubian exodus was particularly dramatic. The evacuees were displaced according to timetables dictated by the dam construction at Aswan and the rising lake, and despite a firm promise made by the Sudanese Prime Minister Abboud that he would accept the choice of resettlement site made by the Nubians themselves, the Government decided to move the population to Khashm el Girba. This was neither the first nor the second choice of the local population which had been presented with a list of six alternatives by the government. The el Girba resettlement represented almost a total break with the past: The Nile with its green banks and islands, covered with a mat of vegetation and with groves of date trees on either side, and surrounded by the vacant expanse of the Sahara with its rainless sands and rocky hills, was left behind and substituted with a flat belt of rainy savannah with a notable lack of trees, hills or anything else that can break the monotony of a flat horizon. Their mode of agriculture was radically altered: With a background in subsistence agriculture and urban careers, they entered a large-scale production organization as lease-holders under a tenancy agreement, to produce crops for the world and the national markets rather than crops for their own subsistence; and, as tenants, they joined this organization along with members of many other ethnic groups; The nomads, on the other hand, who had part of their grazing lands turned into agricultural fields, received little assistance in terms of planned settlement and basic services, and it can be claimed that their settlement was engineered on a purely agricultural basis." (Sørbø 1985, p.12)

A third scholar, Abu Sin, attributed the Scheme failure to the conflict of interests of the planners and the settlers (Davies 1985):

"The wider the cultural and economic gap between communities, the further apart is their perception. Three forms of perception are encountered in the Khashm el Girba Scheme at present: the scheme planners' end management's perception arising from a Western education and urban background, the Nubians' perception arising from a riverain sedentary background, and the tenant nomads' perception derived from a rural nomadic background. Both the Nubians and the nomads come from semi-closed cultural systems which have been exposed to large-scale interaction for the first time in the scheme over the last two decades. The resulting 'perception gap' has been one of the major factors responsible for the deteriorating performance of the scheme.

This gap will be evident if we look into what is meant by development in the minds of each of the groups involved in the scheme. Planners and management, under their perception of what people need, design and manage the scheme on the basis of a simple cost/benefit analysis. The participants at the other end see development as change generated from within, which implies a gradual transformation of the economy starting from what people are pursuing at present. The joint interest in the scheme and a narrowing of the cultural gaps between the parties resulting from increasing contact are expected also to narrow the perception gap between management and participants and among the participants themselves, but at present the parties are still too far apart for any level of flexibility to facilitate coordinate action." (Abu Sin in Davies 1985, p.62)

Another reason for the failure of the Scheme was that there was insufficient water to support and maintain the traditional cropping intensity because of continuous decreases in reservoir storage capacity because of siltation. Water deficiencies disturbed the cropping system, reduced returns and hence tenants' confidence in the scheme. The poor quality of the land in the Butana region is another reason. Butana is really grazing land and not really suitable for cultivation (Salman 2013).

A further reason related to the management, which generally failed to provide the necessary inputs on time, because of such factors as the lack of machinery and spare parts, or fuel shortages. But this in turn was related to Sudan's macroeconomic malaise (Sørbø 1985).

The president of New Halfa farmers union, Awad Elkarim warned of seasonal crop failure because of irrigation problems (Abdulatif 2008). Awad Elkarim said that the state agricultural union had proposed urgent action and its solutions for the crisis, which included prioritising the establishment of the Sitait dam (Abdulatif 2008).

Part 4: Food

4-1: World food situation

Despite that the world produces enough food for a global population of 7.2 billion people[9] (FAO 2002), a total of 842 million people in 2011–13, or around one in eight people in the world, were estimated to be suffering from chronic hunger, regularly not getting enough food to conduct an active life. This figure is lower than the 868 million reported with reference to 2010–12. The total number of undernourished has fallen by 17 percent since1990–92 (FAO 2013). That means, at the turn of this century, world agriculture produced 17 per cent more calories per person than it did 30 years before; despite a 70 per cent population increase (FAO 2002). And yet, many people are suffering from hunger. The list of causes of world hunger and food insecurity is long and multifaceted: they range from macroeconomic imbalances and trade dislocations to environmental degradation, poverty, population growth, gender inequality, inadequate education, poor health and political instability, war and civil strife. All, however, can be related to two basic causes: insufficient national food availability and insufficient access to food by households and individuals (Graßl, Kokott et al. 2003).

New driving factors are currently redefining the world food situation at a rapid pace, transforming food consumption, production and markets. Factors include income growth, climate change, high energy prices, globalisation and urbanisation. Another factor is the private sector, especially the rapidly increasing leverage of food retailers. New producer-consumer links, changes in food availability and rising commodity prices all have crucial implications for the livelihoods of poor and food-insecure people (Von Braun 2007)

A combination of many factors meant that by 2008 global food stocks were at one of the lowest levels on record, having steadily declined since the late 1990s. Many markets in developing countries witnessed sharp increases in staple food costs as a result of significant grain price spikes that occurred in early 2008 (World Bank 2009). In 2011, global cereal production was down by 2.7 per cent and cereal use was projected to decline slightly from the previous season. Severe droughts in 2012 in the USA and across a large part of Europe and into central Asia were the main cause of reduced wheat and coarse grain crop production. However, these reductions would be compensated through an uptake in rice, which would help cereal consumption to remain stable.

Nonetheless, the overall shrinking of grain supplies has increased international grain prices (FAO 2012). According to Luca Chinotti from aid agency Oxfam "lack of political action to tackle high food prices, gender inequality, land grabs

---

9. Enough to provide everyone in the world with at least 2,720 kilocalories (kcal) per person per day (FAO 2002).

and climate change [risks], reversing past gains in the fight against hunger." "The fact that ... more than the population of the US, Europe and Canada are hungry in a world which produces enough for everyone to eat is the biggest scandal of our time," he said (Hornby 2012).

4-2: Food security

The interaction of the factors mentioned above over time results in conditions ranging from acute insecurity (famine), through seasonal discontinuities (lean seasons, climatic shocks), to the loss of guaranteed access and use for all individuals. The concept of food security evolved in the 1990s from a historic focus on the supply of food at the national level (Webb and Rogers 2003). Food security has been defined as "a situation in which all people, at all times, have physical and economic access to sufficient, safe and nutritious food to meet their dietary needs and food preferences for an active healthy life" (World Food Summit and FAO 1996). Three crucial processes produce food insecurity (Mwaniki 2006, Zuberi and Thomas 2012):

1. **Food stocks** (availability) refer to supply of food, determined by nationally produced food plus the food that can be imported to feed the population. In this respect, food supply and its variety should be sufficient in quantity and of good quality (Mwaniki 2006).
2. **Access to food**, determined by the level of poverty – and therefore purchasing power – and transportation and distribution systems, and consumer preferences within a given area.
3. **Food utilization**, corporates aspects such as adequate diet, clean water, sanitation, and health care. It is particularly relevant for integrated food and nutrition security frameworks (FAO/SIFSIA Programme 2005).

A fourth concept is also increasingly becoming accepted: **risks** that can disrupt any one of the first three factors. To achieve food security objectives, all four dimensions must be fulfilled simultaneously.

Food insecurity is a global problem. Some statistics show the magnitude of this problem: almost 1 billion people go hungry in the world; about 25,000 people die every day from hunger-related causes (Sheehy, Ferrer et al. 2008) and one in three children is underweight (Graßl, Kokott et al. 2003). Over the past decade, the number of people suffering from hunger in Africa has increased sharply, whereas in Asia the number has not changed much (see Figure 4). Africa remains the region with the highest prevalence of undernourishment, with more than one in five people estimated to be undernourished. Levels and trends in undernourishment differ within the continent. While sub-Saharan Africa has the highest level of undernourishment, there has been some improvement over the last two decades, with the prevalence of undernourishment declining

Fig 4. Number of people who suffer from hunger

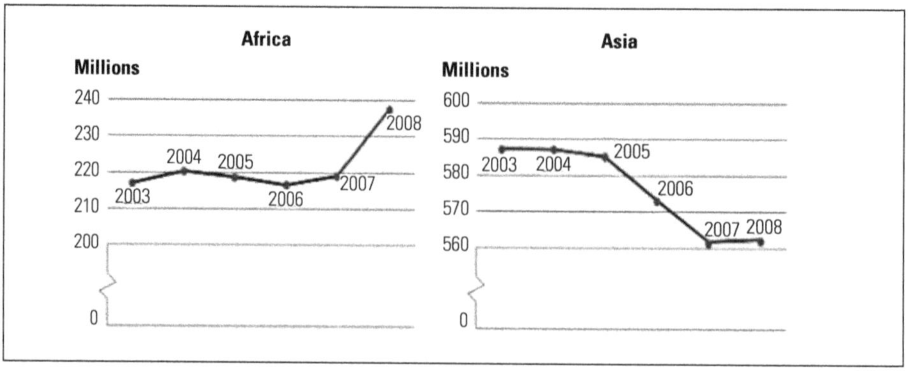

Africa (Millions): 2003 ~215, 2004 ~220, 2005 ~220, 2006 ~215, 2007 ~220, 2008 ~238

Asia (Millions): 2003 ~585, 2004 ~588, 2005 ~590, 2006 ~580, 2007 ~565, 2008 ~565

Source: Food and Agriculture Organization

from 32.7 percent to 24.8 percent ... Both the number and proportion of people undernourished have decreased significantly in most countries in Asia, particularly in South-Eastern Asia, but progress in Southern Asia has been slower, especially in terms of the number of people undernourished. The prevalence of undernourishment is lower in Western Asia than in other parts of the region but has risen steadily since 1990–92. With a decline in prevalence from 31.1 to 10.7 percent, the most rapid progress was recorded in South-Eastern Asia, followed by Eastern Asia (FAO 2013). This was a consequence of the sudden sharp increase in food prices all around the world in 2007 and 2008, and following the economic crisis of 2008–09 (European Court of Auditors 2012).

Developing nations, as well as developed nations, must face the challenge of how to achieve food security in its totality. The difference lies in the magnitude of the problem in terms of its severity and proportion of the population affected (Mwaniki 2006).

Many efforts have been exerted during the past 20 years to reduce the hunger level in the world so as to achieve Millennium Development Goal (MDG) 1 by 2015. The goal was to reduce the hunger by 50% by 2015. Though the situation is getting better, eradication of hunger remains a major global challenge, with almost 870 million people chronically undernourished in 2010–12. This figure of hungry people in the world remains unacceptably high (FAO and IFAD 2012). Out of those 870 million people, 852 million lived in developing countries where the prevalence of undernourishment in 2012 was estimated at 14.9 percent of the population (FAO 2012).

Different rates of progress led to significant changes in the distribution of the undernourished in the world between 1990-92 and 2010–12 (Table 12). Poverty in developing countries is the root cause of food insecurity because it obstructs the ability of people to gain access to food. Whereas many countries

Table 12. Number of undernourished people by region, 1990–92 and 2010–12

Regions	Number of undernourished people (millions)	
	1990–92	2010–12
Developed regions	20	16
Southern Asia	327	304
Sub-Saharan Africa	170	234
Eastern Asia	261	167
Southeast Asia	134	65
Latin America and the Caribbean	65	49
Western Asia	13	25
Caucasus and Central Asia	9	6
Oceania	1	1
Total	**1000**	**868**

Adopted from FAO (All figures are rounded)

have made significant progress towards poverty alleviation – especially Brazil, China and India, albeit to varying degrees and for different reasons (Ravallion 2009) – Africa and sub-Saharan Africa in particular, continues to lag behind.

Africa's malnourished population has almost doubled since the late 1960s. It has almost increased in a similar rate as its population growth. In 2005, one-third of Africa's population, approximately 200 million people, faced hunger and chronic malnutrition. Projections show that the tendency for such events to happen will increase unless preventive measures are taken to avert them. The high prevalence of HIV/AIDS; civil war, conflict and poor governance; frequent drought and famines; and agricultural dependency on the climate and environment are the main factors have contributed to such a tendency (Mwaniki 2006). The continent is constantly threatened by acute food crises and famine. Strategies to reduce poverty and improve food security have hitherto failed (Boussard, Daviron et al. 2005).

There is a high degree of heterogeneity among African countries regarding levels of malnutrition within their populations. In 2005, three sub-Saharan African countries (Gabon, Nigeria and Namibia) showed figures of less than 10% for the prevalence of malnutrition; fewer than half the countries showed figures below 30%. Some countries, despite economic growth and sufficient aggregate availability, display increasing malnutrition, as measured by the prevalence of stunted growth in children (for example, Mali) (FAO 2005).

More than 70 per cent of the food insecure population in Africa lives in rural areas. Table 13 shows the distribution of food insecurity in Africa. Paradoxically, more than 50 per cent of the smallholder farmers and herders, who produce more than 90 per cent of the continent's food supply, account for Africa's food insecure population. (Mwaniki 2006).

Table 13 Proportion of food insecure in Africa

Sub-sector	Food insecure proportion (%)
Farming households	50
Rural landless poor	30
Urban poor	20

Adopted from Heidhues et al. 2004

4-3: Food sovereignty

In the classical model of the food system, food is produced locally to provide the foundation of people's incomes, economies, nutrition, ecologies and culture throughout world. Local food systems provide livelihoods for more than 2.5 billion small-scale farmers, pastoralists, forest dwellers and artisanal fisher folk worldwide (Pimbert 2009). However, despite all of this and the future potential for providing services to people and sustaining diverse ecologies, two main processes threaten local food systems and the organizations that take care of them (Pimbert 2009):

1. The global restructuring of agri-food systems, with a few transnational corporations gaining monopoly control over different links in the food chain. This process is undermining local people's capacity for autonomy and self-determination.
2. The modernist development agenda that organisations such as the World Bank and the Gates Foundation are pursuing. This agenda predicts achieving the MDGs by reducing the number of people engaged in food production and instead encouraging them to get jobs in the largely urban-based manufacturing and service sectors – regardless of the social and ecological costs.

Food sovereignty movement Via Campesina, an international farmers' and peasants' movement, emerged in reaction to this. Food sovereignty is an alternative paradigm for food and agriculture (Pimbert 2009), defined as:

"the right of peoples to define their own food and agriculture; to protect and regulate domestic agricultural production and trade in order to achieve sustainable development objectives; to determine the extent to which they want to be self-reliant; [and] to restrict the dumping of products in their markets… Food sovereignty does not negate trade, but rather, it promotes the formulation of trade policies and practices that serve the rights of peoples to safe, healthy and ecologically sustainable production[10].

The movement called for a Global Rallying Cry of food sovereignty Rome in 2002; it believes that corporate economic globalisation and free trade policies destroy rural communities around the world.

---

10. www.viacampesina.org

The box below explains the difference between the concepts of food security and food sovereignty according to (Rosset 2003):

Box 1. Food security vs food sovereignty

Food security	Food Sovereignty
Every adult and child must have the certainty of having enough to eat each day; it does not specify where that food should come from or how it is produced. Thus, export countries argue that importing cheap food from them is a better way for poor countries to achieve food security than by producing it themselves.	Feeding a nation's people is an issue of national security and sovereignty. If a country's population must depend for its next meal on the vagaries of the global economy; the goodwill of a superpower not to use food as a weapon; or the unpredictability and high cost of long-distance shipping, that country is not secure in terms of either national or food security.

Rosset argues that massive imports of cheap, subsidised food undercut local farmers, driving them off their land. They swell the ranks of the hungry, and their food security is placed in the hands of the cash economy, just as they migrate to urban slums where they cannot find jobs that pay a living wage. Rosset thinks that, to achieve genuine food security, people in rural areas must have access to productive land and receive prices for their crops that allow them to make a decent living (Rosset 2003).

According to Via Campesina and others, farmers and peasants face a clash of economic development models for the rural world. The contrasts between the dominant model – based on agro-exports, free trade, and neoliberal economic policies – and the food sovereignty model is summarised in the box below (Pimbert 2009):

Box 2. Dominant model versus food sovereignty model

Issue	Dominant model	Food sovereignty model
Trade	Free trade in everything	Food and agriculture exempt from trade agreements
Production priority	Agro-exports	Food for local markets
Crop prices	"What the market dictates" (leave intact mechanisms that enforce low prices)	Fair prices that cover costs of production and allow farmers and farmworkers a life with dignity
Market access	Access to foreign markets	Access to local markets; an end to the displacement of farmers from their own markets by agribusiness
Subsidies	Although prohibited in developing countries, many subsidies are allowed in the US and Europe, but are paid only to largest farmers	Subsidies that do not damage other countries (via dumping) are acceptable (e.g. grant subsidies only to family farmers, for direct marketing, price, income support, soil conservation, sustainable farming, research, etc.)
Food	Chiefly a commodity; in practice, this means processed, contaminated food that is full of fat, sugar, high fructose corn syrup and toxic residues	A human right: specifically, should be healthy, nutritious, affordable, culturally appropriate and locally produced
Being able to produce	An option for the economically efficient	A right of rural people
Hunger	Because of low productivity	Distribution; because of poverty and inequality

Food security	Achieved by importing food from where it is cheapest	Greatest when food production is in the hands of the hungry or when food is produced locally
Control over productive resources (land, water, forests)	Privatised	Local; community controlled
Access to land	Via the market	Access to land: the most important issue
Seeds	A patentable commodity	A common heritage of humanity, held in trust by rural communities and cultures; "no patents on life"
Rural credit and investment	From private banks and corporations	From the public sector, designed to support family agriculture
Dumping	Not an issue	Must be prohibited
Monopoly	Not an issue ~	The root of most problems; monopolies must be broken up
Overproduction	No such thing, by definition	Drives prices down and farmers into poverty; we need supply management policies for the US and EU
Genetically modified organisms (GMOs)	The wave of the future	Bad for health and environment; an unnecessary technology
Farming technology	Industrial, monocultural, chemical intensive; uses GMOs	Agro-ecological, sustainable farming methods, no GMOs
Farmers	Anachronistic; the inefficient will disappear	Guardians of culture and crop germplasm; stewards of productive resources; repositories of knowledge; internal markets and building blocks of broad-based, inclusive economic development
Urban consumers	Workers to be paid as little as possible	Need living wages
Another world (alternatives)	Not possible/not of interest	Possible and amply demonstrated (see resources below)

Source: (Pimbert 2009)

4-4: Food situation in Sudan

Periods of drought in the 1970s and early 1980s and more recently have led to reductions in vegetation cover and food production in Sudan. Consequently, they triggered mass migration from farms and villages and thereby upset the economic as well as the social spheres of many areas. Floods and rains in the late 1980s and 1990s were a mixed blessing for both the people and the environment.

Sudan has gone through various political phases marked by (i) civilian rule and multi-party systems; and (ii) military rule and single-party systems. This has affected the country's stability, economic policies and development strategies (Ali 2003) – and its agriculture sector. Political instability, economic strains, which included disruption of natural as well as human resources, and the long civil war in Southern Sudan, should be taken into consideration when discussing the food security situation in Sudan. Conflicts in Darfur, Blue Nile and South Kordofan regions have also affected the food situation in the country.

Economically, Sudan remains a low-income country with a food deficit (IFAD 2009) and does not expect to meet the MDGs by 2015 (Government of Sudan 2008). In 2012, Sudan's Human Development Index (HDI) was 0.414

compared to that of the Arab states as a region with an average HDI of 0.652. Sudan was ranked 171 out of 187 countries and is below the regional average (UNDP 2012).

Decades of conflict, which caused the destruction of physical and human resources, erosion of institutions and damage done to social capital have given rise to a state of poverty in Sudan (FAO 2010). Poverty is presumed to be higher in the rural areas because of low agricultural productivity and high unemployment. High regional inequality, gender disparity/gender income disparity and the existence of a wide range of socio-economic groups characterise the country.

Economic growth triggered by oil industry during last decade had not significantly benefited the poor (IFAD 2009). This may had been because the growth was in sectors such as oil and related businesses, as well as telecommunications, which did not generate sufficient employment for the poor. Other possible factors are that the poor lack secure and fair access to productive assets, in particular land, water and credit; or that they have failed to make immediate use of these opportunities because of under-nutrition, low levels of education, ill health, age or social discrimination. Regional conflict and continuing instability have forced millions of people to flee their homes and left millions more facing extreme poverty (FAO 2010).

In sum, the country experiences high levels of food insecurity because of: the (i) prolonged wars (IFAD 2009) and their consequences on the livelihoods of the people; (ii) recurrent droughts; (iii) floods; and (iv) outbreaks of animal diseases (FAO 2010). Moreover, a dynamic interrelation exists between population, food security, the environment, and natural resources (Zuberi and Thomas 2012).

According to FAO's methodological framework for the global monitoring of prevalence of undernourishment, severity of food insecurity depends on the level of food deprivation, which refers to the condition of people whose food consumption is continuously below a minimum dietary energy requirement. Table 14 below gives the different degrees of the severity of undernourishment (Government of Sudan 2008).

In Sudan, those people whose dietary energy consumption is below the minimum dietary energy requirement (MDER) of 1751 kcals is defined as food

Table 14. Severity of undernourishment (FAO)

Level of food deprivation (%)	Severity of undernourishment
<2.5	Negligible
2.5–4	Very low
5–9	Low
10–19	Moderate
20–34	High
>35	Very high

Source: Central Bureau of Statistics of Sudan, Southern Sudan Centre for Census Statistics Evaluation et al. 2010)

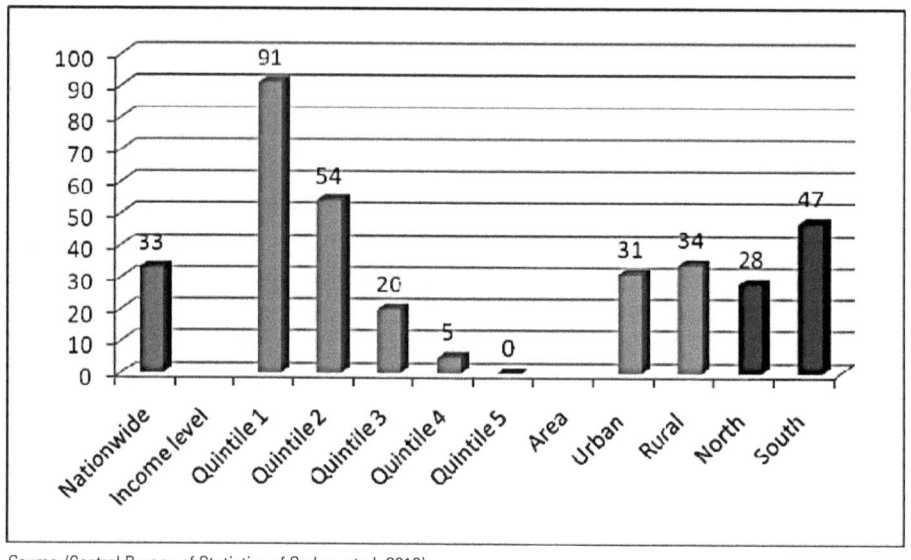

Fig 5. The prevalence of undernourishment in Sudan by income quintile, area and, region

Source: (Central Bureau of Statistics of Sudan, et al. 2010)

deprived. The National Baseline House Survey (NBHS) which was conducted in 2009 found that one in three Sudanese were food deprived, as shown in Figure 5. This places Sudan in the high severity level of undernourishment according to FAO categories illustrated in Table 14.

Other studies on nutritional indicators in Sudan found that 31 per cent of children under the age of five years were underweight. Almost 48 per cent were stunted and 18.1 per cent suffered from moderate or severe/acute malnutrition (SHHS, 2006 (Government of Sudan 2008)). The prevalence of undernourishment was 31 per cent and 34 per cent in urban and rural populations.

In terms of the prevalence of undernourishment, the northern states were categorised as experiencing high-severity food deprivation, and 47% of southern states were experiencing very high-severity food deprivation. The highest levels of food deprivation were observed in the states of Western Bahr al Ghazal, Unity, Upper Nile, Warrap and Lakes, affecting more than half of their populations (shown in black in Map 6).

In 2009, Sudan remained the largest single recipient for the fifth consecutive year, with US$1.4 billion. Despite this, millions of people continue to face severe and chronic food insecurity (FAO 2010). For decades, many parts of Sudan have suffered frequent periods of acute food insecurity, as well as chronic food insecurity. In Darfur, the Nuba Mountains and Blue Nile State the causes are mainly related to conflict, but are also the result of natural disasters, resource degradation and mismanagement.

Map 6. Food and nutrition security assessment in the Sudan

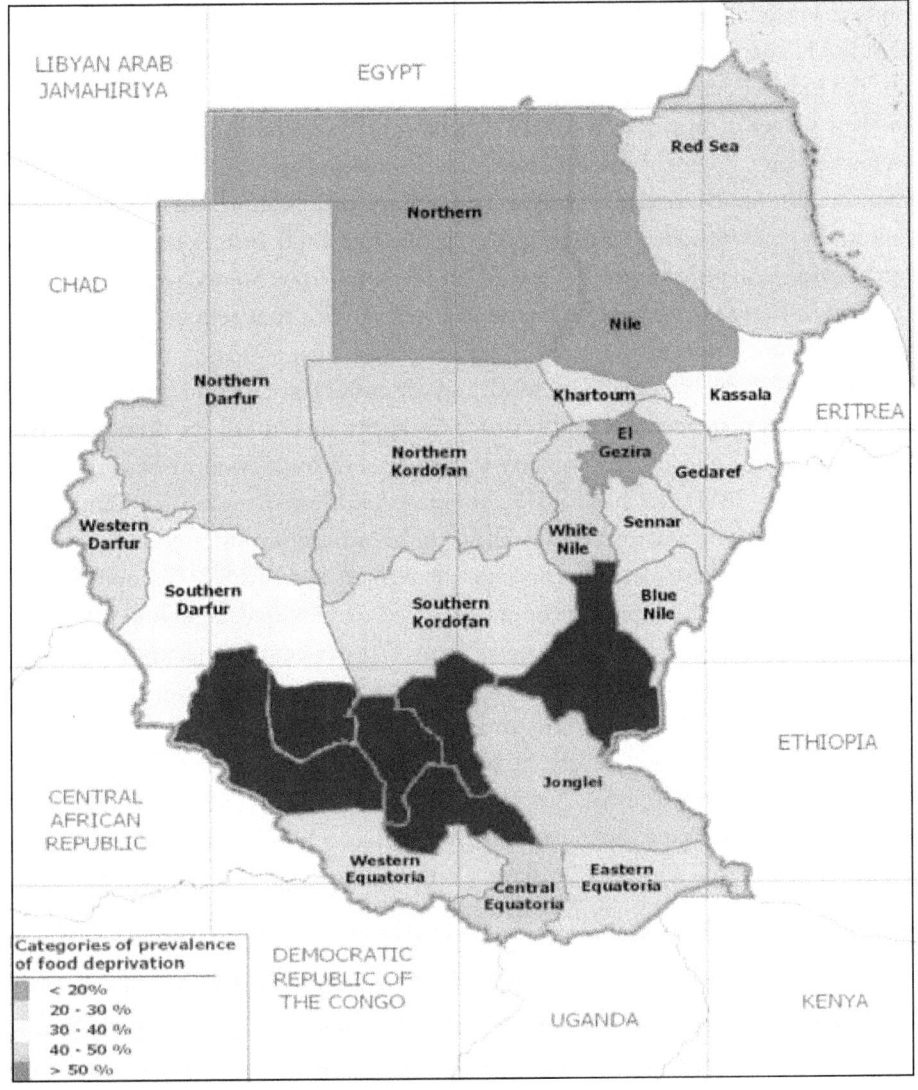

Source: (Central Bureau of Statistics of Sudan, et al. 2010)

The protracted crisis that affected many rural people in northern and eastern regions results from environmental degradation, the increasing incidence of drought and chronic poverty caused by long-term political and economic marginalisation. The result has been a more or less permanent state of severe food insecurity, with alarming declines in measures needed to promote human well-being, such as health and nutrition. An estimated 20 per cent of the nearly

37 million people in Sudan are chronically undernourished (FAO 2010). The highest levels of poverty and food insecurity are recorded among traditional, rain-fed farmers and pastoralists (World Bank, 2003) (FAO 2010).

The independence of South Sudan and a surge in refugees and IDPs returning to their places of origin marked 2011. The most vulnerable and insecure areas were the Three Protocol Areas (Abyei, Southern Kordofan and Blue Nile States) that border South Sudan, Darfur and Khartoum State (FAO 2011). Many factors have entrenched chronic poverty, such as high inflation, depreciation of the Sudanese pound and the loss of oil revenues from South Sudan, political turmoil in parts of the Middle East and North Africa that reduced remittances from Sudanese migrants.

Sudan is a net importer of food and essential agricultural inputs. Along with conflict, displacement and reliance on irregular rainfall for domestic crop production, this has left the country in a state of chronic food insecurity (FAO in emergency 2011). In June 2012, about 4.7 million people in Sudan faced differing levels of food insecurity. This was because of conflict and civil strife, poor 2011-12 harvests, macroeconomic instability, and severely disrupted trade flows that have limited market supplies and led to above-average food prices. In May 2012, sorghum prices were more than 170 per cent higher than the five-year average and about 90 per cent higher than the reference year (2009/2010) (USAID and NET 2012, USAID and NET 2012).

The impacts are most severe in border areas, where conflict, displacement and trade restrictions prevail. Since the secession of South Sudan in July 2011 inflation rates have continued to soar. According to Sudan's Central Bureau of Statistics (CBS), the May 2012 inflation rate was 30.4 per cent compared to 28.6 per cent in April. The government recently instituted several major economic reforms, which included the gradual removal of fuel subsidies.

This removal of fuel subsidies is challenging because it will sharply increase the cost of living, transport, and agricultural products (USAID and NET 2012). Furthermore, because of a poor rainy season between 2011 and 2012, and reduced purchasing power because of high inflation, approximately 1 million people Red Sea, North Kordofan, White Nile and Kassala States have been susceptible to severe levels of food insecurity. Additionally, approximately 100,000-120,000 IDPs in the Abyei area and the majority of Darfur's 1.8 million IDPs face stressed levels of food insecurity (USAID 2012).

During August-September 2012, above-average rains caused widespread flash floods in many parts of Sudan. About 100,000 people were affected. The floods caused damage to houses and property, killed thousands of animals and destroyed thousands of hectares of cropland (USAID 2012). In October 2012, it also led to acute food insecurity for an estimated 3.2–3.5 million people. However, this figure is about 30%-35% lower than the figure of 4.6 million

Map 7 Estimated food security outcomes, October 2012

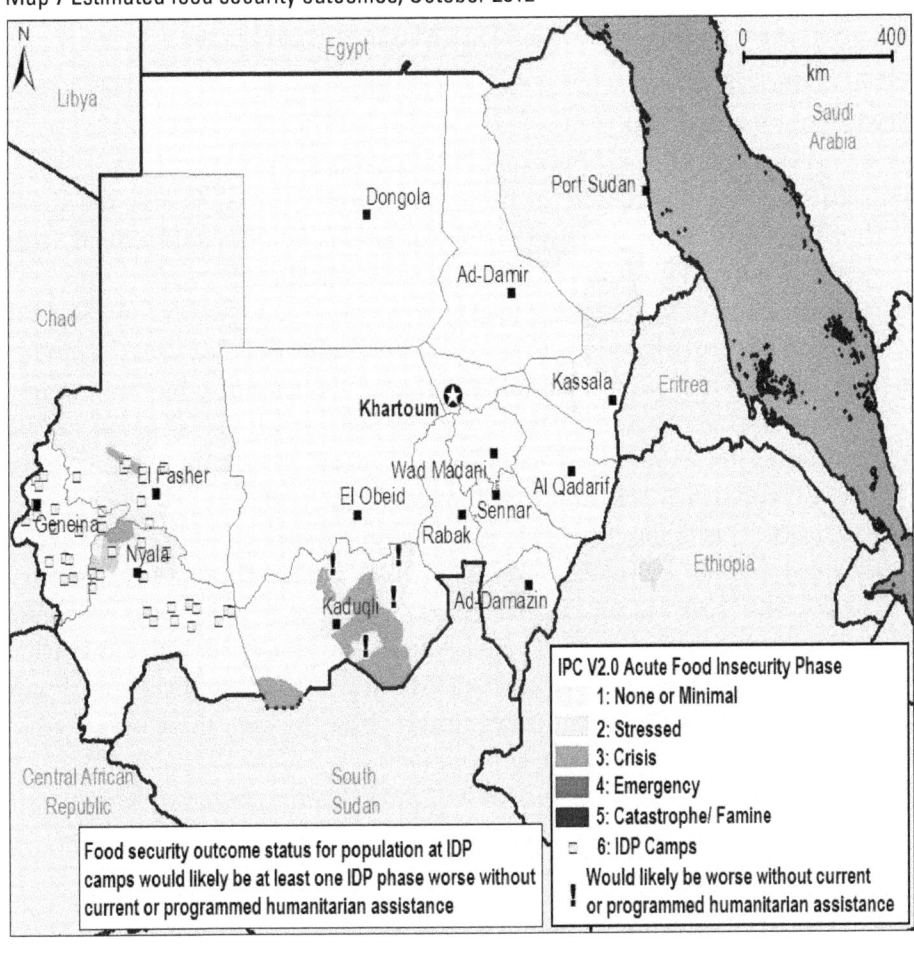

Source: (Famine Early Warning Systems Network 2012)

people who experienced food problems in July 2012 (Famine Early Warning Systems Network 2012).

4-4: Causes of food insecurity in Sudan

The different conflicts, war and the counter-insurgency warfare are considered to be the main causes of food insecurity in Sudan, particularly in the south and some northern states. They have had a direct impact on food security through:
- Destruction of infrastructure.
- Damage to the social and economic fabric.
- Weakening institutions that provide services.
- Mass population displacement.

- Warring factions preventing international assistance and food relief.
- Government policies regarding Taxes, Marketing and Finance.

(Central Bureau of Statistics of Sudan, Southern Sudan Centre for Census Statistics Evaluation et al. 2010)

Other compounding factors can also be noted such as:

- Recurrent natural disasters in the form of drought and floods; the worst droughts occurred in 1983-84, 1997-98, and 2000-2001 and caused large population displacements and high livestock mortality.
- Limited access to basic services such as water supply and health services. This has been particularly serious in some states, notably Red Sea (IFAD 2009).
- Lack of infrastructure has negatively affected food security by limiting free movement of goods and food for the internal market. Food could not be transported from food-producing regions, or areas with food surpluses to areas of food deficit, which resulted in escalating food prices;
- Insecurity in neighbouring countries has caused a large influx of refugees. Coupled with an estimated 5.8 million IDPs scattered around the country (FAO 2010), this has put a further strain on already poor socio-economic and environmental resources. In addition, large population movements around the country occurred after the signing of the Comprehensive Peace Agreement in 2005 and the referendum in 2011. In sum, these have further added strains to the food security situation.
- Finally, shortcomings in the policy and programming framework, and poor institutional set-ups have not been conducive to improving food security.

Part 5: Climate

5-1: Agriculture and global food security under climate change

According to the Intergovernmental Panel on Climate Change report (eds Houghton et al. 1990), atmospheric carbon dioxide (CO_2) is expected to increase from present day concentrations of about 350μL/L to more than 800μL/L by the end of the century if no steps are taken to limit emissions. This increase in CO_2 concentration in the atmosphere leads climate modellers and experts to predict a consequent global warming and changes in precipitation patterns. Additionally, it is projected that this increase in CO_2 and possible associated climate change could affect the ecology of most living things, including agricultural production.

The IPCC has defined climate variability as "the variations in the mean state and other statistics (such as standard deviations, statistics of extremes, etc.) of the climate on all temporal and spatial scales beyond that of individual weather events" (IPCC 2007). Variability may be caused by natural internal processes within the climate system (internal variability), or to variations in natural or anthropogenic external forces (external variability). Therefore, the IPCC defines climate change as "any change in climate over time, whether due to natural variability or as a result of human activity" (IPCC in Elasha 2010 p.11).

According to the IPCC reports (2007), the global surface air temperature rose by 0.76°C from 1850 to 2005. The IPCC observed that the rise in air temperature followed a linear trend over the previous 50 years. This increase in temperature has led to widespread melting of snow and ice and a rising global mean sea level. The same report also predicts an increase in the magnitude and frequency of extreme weather events, such as storms, precipitation and drought; and more intense tropical cyclones (hurricanes and typhoons), characterised by higher peak wind speeds; and heavier precipitation associated with warmer tropical seas.

Climate change may affect agriculture and food production systems in complex ways, directly and indirectly. Directly, for example, climate-induced changes in agro-ecological conditions are likely to affect crop production (e.g. changes in rainfall leading to drought or flooding, or warmer or cooler temperatures leading to changes in the length of growing seasons). The US Department of Agriculture's Economic Research Service (ERS) anticipates that global warming will shorten growing seasons in the tropics and lengthen growing seasons at high latitudes, whereas the impact on mid-latitude zones will be mixed. The magnitude of these effects increases as mean global temperature increases (Darwin 2001).

Indirect effects of climate-induced changes on agriculture and food production systems will be through changes in markets, food prices and supply chain infrastructure, which will affect the growth and distribution of income, and thus demand for agricultural produce (Gregory, Ingram et al. 2005, Schmidhuber and Tubiello 2007). However, the effects of higher temperatures are expected to be an advantage to countries in temperate latitudes: the length of the growing period will increase; areas potentially suitable for cropping will expand; and crop yields may rise.

Grassland in some humid and temperate zones may benefit from moderate warming because pasture productivity may increase and hence reduce the need for livestock housing and compound feed. Conversely, higher temperatures may disadvantage other regions. For example, the Mediterranean region may experience heat waves and droughts; increased heavy precipitation events and flooding may occur in temperate regions, including the possibility of increased coastal storms; and semi-arid and arid pastures are likely to see reduced livestock productivity and increased mortality.

Climate models predict increased evapotranspiration and lower soil moisture levels in drier areas. As a result, tropical grassland may adversely affected and become increasingly arid, and cultivated areas may become unsuitable for cropping. Furthermore, rising temperatures will also expand the range of many agricultural pests and increase the ability of pest populations to survive the winter and attack spring crops (Schmidhuber and Tubiello 2007).

The increase in atmospheric CO_2 concentrations will affect agriculture in other ways. Plants generally exhibit faster growth when the CO_2 around their leaves is increased (Kimball, Mauney et al. 1993). Higher levels of CO_2 will improve water-use efficiency of all crops (by reducing evapotranspiration) and increase the rate of photosynthesis (Darwin 2001). Studies on greenhouses and growth chambers show that plant growth and yields have typically increased by more than 30 per cent with a doubling of CO_2 concentrations.

Interactions between CO_2 and climate variables also appear important; other studies suggest that the combination of higher temperatures and increased levels of CO_2 may produce more than a 30 per cent increase in growth rates and yields. Several studies have suggested that during periods of water stress, CO_2 growth stimulation is more significant than under well-watered conditions. Therefore, the direct CO_2 effect will to some extent compensate for a hotter drier climate.

However, an increase in yields does not necessarily mean an increase in the nutritional quality of agricultural produce. Some cereal and forage crops, for example, show lower protein concentrations under elevated CO_2 conditions (Schmidhuber and Tubiello 2007).

The direct effects of high concentrations of CO_2, will be insignificant in regions where other factors inhibit crop growth or low fertiliser use. Even if higher concentrations of CO_2 benefit some regions in the world, the direct detrimental effects of other fossil fuel emissions, such as sulphur dioxide and ozone, will offset some of these benefits (Darwin 2001). Furthermore, the role of agriculture in climate change should not be ignored. Agriculture is a major contributor of the greenhouse gases methane (CH_4) and nitrous oxide (N_2O), and regionally derived policies that promote adapted food systems need to mitigate further climate change (Gregory, Ingram et al. 2005).

5-2: Impacts of climate change in Africa

The African continent has a land area of nearly 30 million km^2. It is endowed with many natural resources that are largely unexploited and that are found in very few parts of the world. These include minerals, forests, wildlife and rich biological diversity. However, this natural wealth is not reflected in welfare measures directed at the region's inhabitants (Urama and Ozor 2010). The climate ranges from Mediterranean-type climates through seasonally arid tropical to humid equatorial.

Annual precipitation across the continent averages 678 mm. The northern region, which covers about 20 per cent of the total area, receives less than 3 per cent of the total precipitation. In contrast, the central region, which has a similar area, receives 37 per cent of all precipitation in Africa (Frenken 2005). Temperatures in Africa are high throughout the year. The diurnal range is about 10°C–15°C, except in the deserts (Moges, Awulachew et al. 2011).

Despite the abundance of natural resources, Africa includes some of the world's poorest nations, which makes it particularly susceptible to climate change because of chronic poverty and also because of multiple stresses and low adaptive capacity (Strzepek and McCluskey 2007). One of the most significant impacts of climate change is likely to be on the hydrological system. This will be most evident in persistent signs of decrease in water resources such as lakes and rivers, snow cover on the high mountains, precipitation and water vapour pattern distortions, and poor water quality in surface and groundwater systems.

This will be particularly true in arid and semi-arid areas of Africa, where water resources are very sensitive to climate variability, particularly rainfall (Strzepek and McCluskey 2007). By 2020, between 75 million and 250 million people across the continent are projected to be exposed to increased water stress because of climate change. This situation will worsen if it is coupled with increased demand for water, which will adversely affect livelihoods and exacerbate water-related problems (Urama and Ozor 2010).

Erratic rainfall with high spatial and temporal variability, as well as high evaporation rates, greatly affects agricultural systems from which 70 per cent of the continent's population derive their livelihoods (Strzepek and McCluskey 2007). Projected sea-level rise towards the end of the 21st century, will affect low-lying coastal areas with large populations. Adaptation measures cost could amount to at least 5%–10% of GDP. Mangroves and coral reefs are projected to be further degraded, with additional consequences for fisheries and tourism (Reid 2009).

The future impact of climate change on the water resources in Africa remains uncertain, but it is likely that many regions will suffer from droughts and floods with greater frequency and intensity, if with uncertain regularity. Given that 90 per cent of water resources in Africa are trans-boundary in nature, it is imperative to underline the importance of regional cooperation in all planning pertaining to climate change (Moges, Awulachew et al. 2011).

It is projected that climate variability and change will adversely affect agricultural systems and access to food in many African countries and regions. This will be manifested in decreases in the length of growing seasons and yield potential, and the area suitable for agriculture, particularly along the margins of semi-arid and arid areas. This in turn would adversely affect food security and exacerbate malnutrition in the continent.

There is a risk that yields from rain-fed agriculture could be reduced by up to 50 per cent in some countries by 2020 (Reid 2009). A study by Ringler et al. (2010) concluded that cereal production in sub-Saharan Africa is projected to decline by 3.2 per cent as a result of climate change, with declines in yield of 4.6 per cent, partially compensated by a 2.1 percent increase in area. Among staple crops, negative yield impacts are projected to be largest for wheat, followed by sweet potato, whereas overall yields for millet and sorghum are projected to be slightly higher under climate change. Maize, rice, and wheat prices are expected to increase by 4 per cent, 7 per cent, and 15 per cent by 2050 (Ringler, Zhu et al. 2010).

Local food supplies are projected to be negatively affected by decreasing fisheries resources in large lakes because of rising water temperatures, compounded by continued over-fishing, which will exacerbate and worsen the situation (Reid 2009). Childhood malnutrition levels are projected to increase as a result of climate change across sub-Saharan Africa, with incremental increases from climate change alone of just below 1 million children by 2030. By 2050, 585,000 children will still be malnourished (Ringler, Zhu et al. 2010). Some adaptation to current climate variability is taking place; however, this may be insufficient for future changes in climate (Strzepek and McCluskey 2007).

5-3: Impacts of climate change in Sudan

Sudan is one of the least-developed countries in Africa and one of the most vulnerable states to climate change and climate variability, because majority of its land is arid and desert (Elasha, Elhassan et al. 2005). Furthermore, the country's major economic sector is vulnerable to climate change. Challenges that exacerbate this situation include: endemic poverty; the instable political situation and corruption; a lack of systematic managerial institutions; limited access to capital, including markets, infrastructure and technology; ecosystem degradation; and complex disasters and conflicts. These in turn have negatively affected the population and weakened their adaptive capacity, hence, increasing their vulnerability to projected climate change (Boko 2008).

Sudan's inherent vulnerability may best be captured by the fact that food security is mainly determined by rainfall, particularly in rural areas where more than 65% of the population live (Nimir and Elgizouli 2011). Rainfall is erratic and varies significantly from north to south. The unreliable nature of rainfall, together with its concentration during the short growing season, increases the vulnerability of the rain-fed agricultural system.

A trend of decreasing annual rainfall over the past 60 years (0.5%) and increased rainfall variability is contributing to drought conditions in many parts of the country. This pattern has led to serious and prolonged drought episodes. For example, Sudan experienced a succession of dry years from 1978 to 1987 that resulted in severe social and economic impacts, including the death of people and livestock, and displacement of several million people. Drought problems such as these will increase if the trend continues (Nimir and Elgizouli 2011).

It is projected that the average temperature will rise considerably compared to baseline expectations. Increases in the temperature by 2060 will be in the range from 1.5°C–3.1°C in August to 1.1°C–2.1°C in January. Models predict an average rainfall decrease of about 6 mm/month during the rainy season; hence, traditional rain-fed farmers and pastoralists would be the most vulnerable groups (Elasha, Elhassan et al. 2005).

According to a United Nations Environment Programme (UNEP) assessment, Sudan along with other countries in the Sahel belt, has suffered several long and devastating droughts in the past few decades. The most severe drought occurred in 1980-84, and was accompanied by widespread displacement and localised famine. Localised and less severe droughts (affecting between one and five states) were also recorded in 1967-73, 1987, 1989, 1990, 1991, 1993 and 2000. These drought episodes were in addition to the erosion of natural resources as a result of climate change, which is among the root causes of social strife and conflict.

The scale of historic climate change, as recorded in Northern Darfur, is almost unprecedented: the reduction in rainfall has turned millions of hectares of already marginal semi-desert grazing land into desert (IRINAfrica 2007). Besides droughts, Sudan's National Adaptation Program of Action (Government of Sudan 2007) identified extreme flooding events as the major climate-related hazards associated with climate change.

Two major types of flood event regularly affect Sudan. The first occurs during torrential rains, when high levels of water overflow the Nile River and its tributaries. This type of flood occurs mainly during the rainy season, as

Map 8. Areas most vulnerable to climate change in Sudan

Source: (Zakieldeen 2009: Courtesy by Alasha)

happened in 1946, 1988, 1994, 1998, 1999 and 2001. The other type is flash-flooding, which results from heavy localised rainfall during the rainy summer season or over the Red Sea area in winter season because of mountain run-off. This occurred in 1952, 1962–65, 1978–79, and 1997 (Government of Sudan 2007).

The table below summarises extreme weather and climate events.

Table 15 Extreme weather and climate events in Sudan

Event	Frequency	Vulnerable areas	Sectors affected	Impacts
Drought	Frequent	North and west (North Kordofan, Darfur); Kassala State; some central rain-fed areas	Agriculture; livestock; water resources; and health	Loss of crops and livestock (food shortage); decline in hydropower; population displacement; wildfires
Floods	Frequent	Nile basin and low-lying areas from extreme south to far north; mountain areas along Red Sea	Agriculture; livestock; water resources; health	Loss of life, crops, and livestock; insects and plant diseases; decline in hydropower; damage to infrastructure and settlement areas
Dust storms	Frequent	Centre and north	Transport (aviation and land traffic)	Air and land traffic accidents; health
Thunder storms	Infrequent	Rain-fed areas throughout Sudan	Aviation	Loss of crops, livestock and property
Heat waves	Rare	Northern and central areas; Red Sea State	Agriculture; livestock; health	Loss of life, livestock and crops
Wind storms	Rare	Central and north-central Sudan	Settlements; service infrastructure	Loss of life and property; damage to infrastructure (electricity and telephone lines)

Source: Government of Sudan 2007

Sudan's First National Communication to the United Nations Framework Convention on Climate Change (UNFCCC) and National Adaptation Programmes of Action (NAPA) identified the three sectors most vulnerable to climate change: water resources; agriculture (food production and economic livelihoods); and human health.

They are intrinsically linked. An enormous amount of water is required to produce food under the evolving conditions of climate change. Achieving food security and water security throughout Sudan are major challenges, with the background of changing climate conditions, especially in relation to population. Though population density in Sudan is about10 people/km^2, this density is considerably higher on arable land (63/km^2) and higher still on cultivated land (370/km^2). This means much of the population is clustered in central Sudan and along the Nile River (Government of Sudan 2007).

The Blue Nile is an important trans-boundary source of water, which Ethiopia, Sudan and, Egypt share. In contrast to Sudan and Egypt, Ethiopia has used very little Nile water. However, with a view to driving economic development, the Ethiopian government intends to utilise Nile water resources for irrigation and hydropower; it plans to build many large dams – indeed, work has already begun on some – on the main stem of the river and on tributaries. If all the schemes planned are implemented, the water stored in large reservoirs will exceed 160,000 billion m^3 (i.e. approximately x14 present levels and x3 the current mean annual flow at the Ethiopia-Sudan border); areas to be irrigated will exceed 360,000 ha (x23 present levels); and installed hydropower generating capacity will be in excess of 10,000 MW (x47 present levels) (McCartney and Girma 2012).

McCartney et al. have identified and modelled four regional scenarios. Their conclusion is that changes in climate will affect water availability and demand. In the first scenario, in a natural situation (i.e. no development), flows would increase slightly at the Ethiopia-Sudan border in the first half of the 21st century, but decrease by 20 per cent, in the second half of the century. Average annual demand for irrigation water would fluctuate but increase significantly, particularly in the second half of the century.

The planned water resources development in the basin would cause an additional 2.6 per cent decline in flows at the border in the second half of the century. However, although there remains great uncertainty about how climate change will affect the water resources of the basin, it is clear that the changes would be likely to have serious consequences for economic development, food security and poverty in the region.

The impacts of a harsher climate change – which based on current emissions trends is equally and perhaps even more likely – would be even more severe (McCartney 2012). Given that 85 per cent of the Nile's water flows from the Ethiopian highlands, climate change compounded by rapid population growth would increase competition over water in the region. Some models found a tendency towards lower Nile flows in all eight of the climate scenarios, with impacts ranging from no change to around a 40 per cent reduction in flows by 2025, to a reduction of more than 60 per cent by 2050 in three of the flow scenarios (Elasha 2010).

Water resources in Sudan are expected be affected by climate change through decreased precipitation and/or increased temperatures and evaporation, which reduce groundwater recharge. Moreover, soil moisture is also likely to decline under future climate change. The overall result of decreased water resources, coupled with other factors such as increased water consumption, population growth and high rainfall variability is that the country could face a serious water crisis (Government of Sudan 2003, Government of Sudan 2007).

Globally, considerable attention has been given to climate change and its impacts on agriculture, because it considered to be one of the sectors that most vulnerable to climate change, and is also an important sector for international trade. In low-latitude regions, where most developing countries are located, reductions of about 5%–10% in the yields of major cereal crops are projected (Tamiotti, World Trade et al. 2009). Moreover, a climate sensitivity analysis of agriculture concluded that some African countries will lose virtually their entire rain-fed agriculture by 2100 (Mendelsohn et al. 2000).

Agriculture (food production) in Sudan faces enormous challenges. Climatic

Fig 6. Rainfall Patterns in Sudan

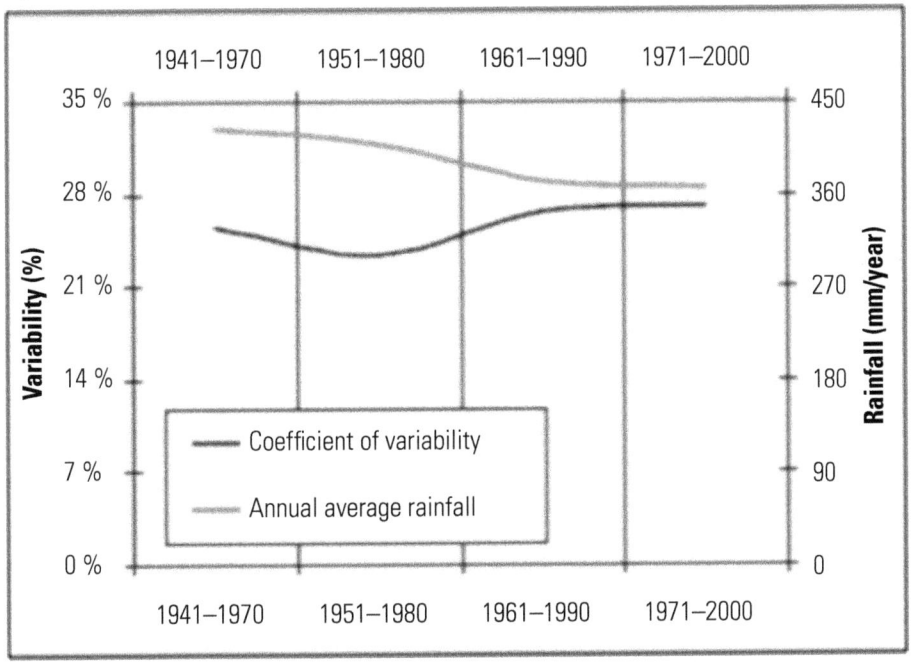

Source: (Government of Sudan 2007: Courtsey by Alasha)

changes are affecting agriculture through their direct and indirect impact on crops, soils, and pests, which in turn affect food availability as a basic pillar of sustainable food security. Therefore, achievement of food security and managing the adverse impacts of climate change are essential. Sudan's agriculture mainly depends on rainfall. A study by Elgali et al. (2010) showed a considerable decrease in cereal production, especially sorghum and millet – which are rain fed – compared to imported crops of wheat and rice (Elgali, Mustafa et al. 2010).

The value of import substitutes shows a remarkable increase to compensate the loss of sorghum and millet. Export crops of sesame, groundnuts, gum Arabic and livestock, all show a considerable decrease, which is reflected in their low export value. The result of the increase in imports and decrease in exports is shown by a deficit in agricultural balance of trade in Sudan as a result of climate change (Elgali, Mustafa et al. 2010).

In conclusion, combined with growing socioeconomic pressures, the imposition of climate variability and climate change is likely to intensify the ongoing process of desertification of arable areas. Humid agro-climatic zones will shift southwards, rendering northern areas increasingly unsuitable for agriculture. Crop production is predicted to decline substantially for both millet and sorghum (see figure 7). The area of arable land, as well as the important gum

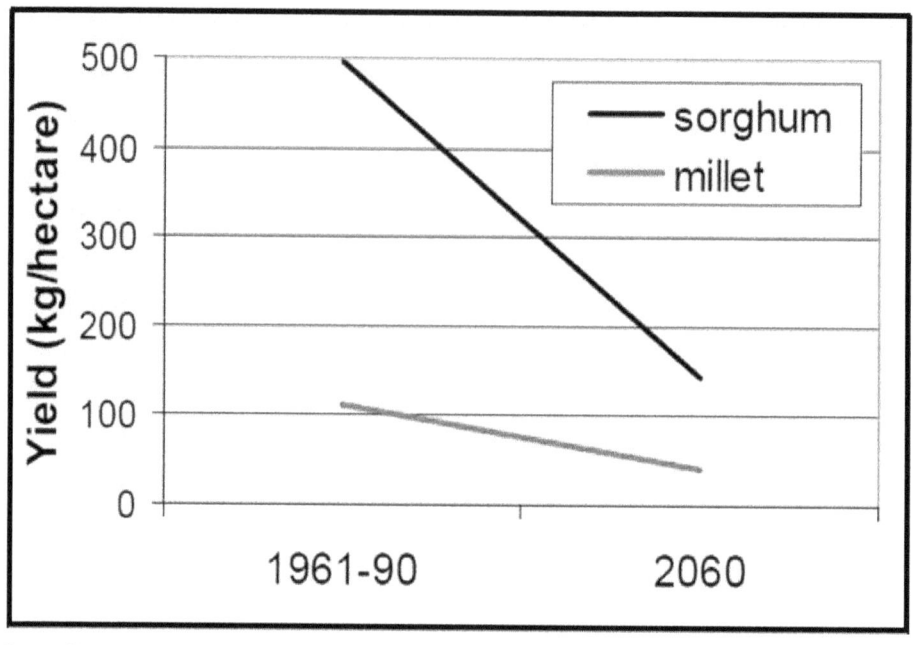

Fig 7. Projected agricultural yields in Sudan with climate change

Source: (Government of Sudan 2007: Courtsey by Alasha)

Arabic belt, would also be likely to decrease, with attendant impacts on local incomes and food security (Government of Sudan 2007).

The effect of climate change of food security is great because it drives vulnerable groups of people into hunger and malnutrition. Droughts, flooding, pest and pathogen outbreaks lead to a loss of agricultural production in terms of quality and quantity. Food insecurity and the resulting poor nutrition status, which increases vulnerability to climate-related diseases, leads to decreased food production because of low labour productivity (Maeng 2012).

According to a report by DARA, 225,000 people die each year from hunger caused by climate change, and this figure will to increase to nearly 400,000 if effective measures are not taken. The report also points out that the least independent groups, such as infants and children, and communities in low-income and the least-developed countries, are hardest hit (Dara Climate Vulnerable Forum 2012). In Sudan, the population could be exposed to a significantly increased risk of malaria. Studies in Kordofan State, for example, have shown that the risk of transmission potential could increase substantially by 2060 (Figure 8) (Government of Sudan 2007).

If this happens, not only would the overburdened health care system experience extreme stress, but the disease would exact a heavy toll on local communities (Government of Sudan 2007).

Fig 8. Malaria transmission in 2060 relative to baseline in Sudan

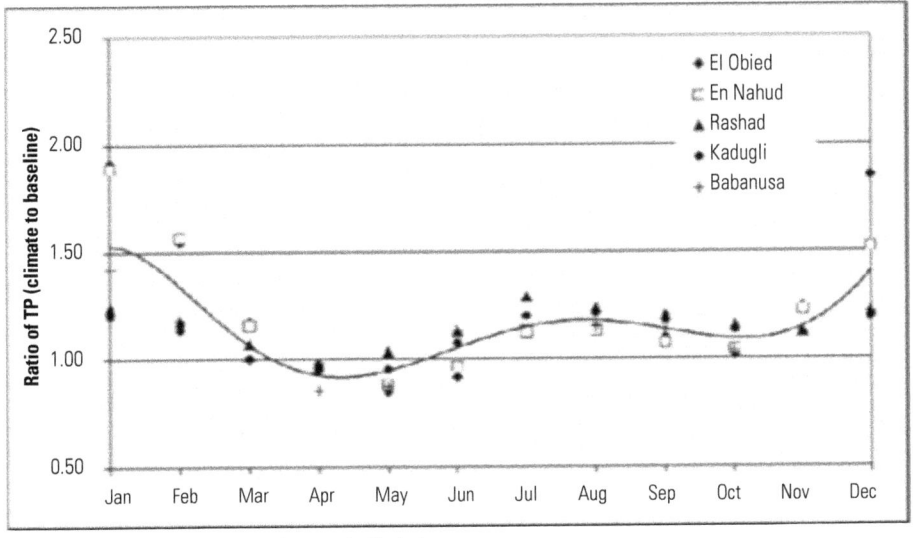

Source: (Government of Sudan 2007: Courtsey by Alasha)

Climate-induced resource conflicts

The UN Security Council has recognised climate change as 'a threat to world peace' by UN Security Council. Competition over natural resources, especially land, has become an issue of major concern and cause of conflict among the pastoral and farming populations across the Sahel and the Horn of Africa. Sudan, where pastoralists constitute more than 20 per cent of the population, is no exception (UNDP 2006). UNEP identified climate change (including deforestation and desertification, etc.) as a major cause of conflict in Sudan, and directly related to the conflict in Darfur region, where desertification has added significantly to the stress on the livelihoods of pastoralist societies, forcing them to move south to find pasture (UNEP 2007).

The history of Sudan since its independence in 1956 has been dominated by long, recurring, and bloody civil wars, which have been attributed either to an age-old racial and ethnic divide between Arabs and Africans or to colonially constructed inequalities. Civil wars have directly affected more than 60 per cent of the country (Johnson 2003). At the same time, Sudan suffers from a number of severe environmental problems, both within and outside current and historical conflict-affected areas.

Concerning the chronology and geography of the various confrontations, tribal and small-scale conflicts fought only with small arms have occurred continuously – over local politics, access to water and grazing, and cattle theft – all over the country, with the exception of the north. However, the majority of large-scale conflicts that have extended over five years or more have been confrontations between forces aligned with the central government based in Khartoum and an array of anti-government forces either in the south, centre or east of the country. The death toll is unknown, but a range of sources estimate it to be in the range of 2-3 million (UNEP 2007).

Two issues have driven violence in Darfur, one local, the other national. The local grievance focused on land; its deep background was a colonial legacy of parcelling Darfur between tribes, with some given homelands and others not; but its immediate background was a process of drought and desertification that had extended across four decades and exacerbated the conflict between the tribes. The national context was a rebellion that brought the state into an ongoing civil (tribal) war (Mamdani 2009).

The conflict in Darfur began as a localised conflict (1987-89) and turned into a rebellion (beginning in 2003). The tribes with land sought to keep out landless or land-poor tribes who were fleeing advancing drought and desert. As early as 1989 at a reconciliation conference in Darfur, the land-owning tribes were already using the language of "genocide" – and indeed "holocaust." – but they made the charge against the coalition of tribes they fought, and not against the government of Sudan (Mamdani 2009).

5-4: Climate change adaptation measures

Many options are suggested worldwide to farmers and governments to counter global warming. Some of these are (Darwin 2001):

- Switching crop varieties.
- Introducing more suitable crops
- Shifting from crops to grazing.

Governments could:

- Provide reliable long-term weather forecasts or information about suitable crop and livestock alternatives to help farmers increase production efficiency.
- Introduce of new varieties better able to withstand the effects of global warming.
- Provide irrigation or/and increase its efficiency.
- Maintain flood control, which requires long-term cooperation with farmers or other members of society.
- Relocate agricultural production. This applies in those cases where government policies have provided reliable, long-term information that identifies suitable and unsuitable crop locations as the climate changes.
- Adopt policies that facilitate the migration of people from one location to another or transition from one profession to another.
- Adopt policies that stimulate economic growth and development and thereby provide more alternatives to agriculture as a source of livelihood.

These options would benefit from reliable, long-term information about climate change and its effects on land and water resources, and consequently on agricultural production and food security.

5-5: Conclusion

After more than half a century, Sudan is still suffering from the political, economic and social scars that colonial rule left behind. The country has engaged in two long-lasting civil wars with what is now the independent South Sudan. Conflicts continue in many parts of the country. This state of unrest and instability has taken its toll on country's human and natural resources, and led to chronic poverty that affects the overwhelming majority of the population.

Sudan is one of the least-developed countries in Africa. Making rational use of its huge and diverse resources is an unfulfilled promise and outstanding challenge. Moreover, development projects have either never seen physical progress or have always been poor. The situation is getting worse because of the economic crisis. Late delivery of necessary equipment and supplies because of transportation difficulties affects most projects. In addition, the continuing drain of skilled labour to neighbouring countries, where job opportunities are better, takes its toll on effective preparation and implementation of development projects.

In relation to agriculture, one can argue, however, that the Sudanese state since the early days of independence has failed to properly and fairly utilise the country's agricultural resources. This has transformed the country into a land of poverty and environmental degradation. Nevertheless, agriculture is the mainstay of the economy. It contributes about 41 per cent of GDP and 80 per cent of exports; it employs more than 65 per cent of the labour force and provides 50 per cent of raw materials for the industrial sector.

Although these contributions seem ample, they are considered small when compared to the potential of the sector. This modest performance has many reasons, which some relate to weather conditions; and others to distorted agricultural policies and inadequate investment in agriculture, which can be seen in poor infrastructure, the low technical capacity of labour force, poor support services (including research and development), and shortages of necessary inputs, such as fertilisers, machinery and so on.

The centrepiece of Sudan's agricultural development strategy is its irrigated schemes. Complex modern gravity-irrigated schemes in sparsely settled parts of central Sudan have become a distinguishing feature of contemporary agriculture in the country. These schemes are identical in design and objectives, and their implementation is poorly coordinated, with little regard to perceptions of development or socioeconomic impact. This neglect results in serious conflicts between the management boards responsible for implementation and the settlers or participants who are to be integrated into the scheme. This is one of the major causes of the ongoing decline in productivity of all major agricultural schemes in the country's semi-arid areas.

The role that these schemes could play in avoiding famine and their potential to increase the national food supply has received little attention from government and/or researchers. Consequently, the irrigated schemes themselves are in a food crisis that has resulted from inconsistent agricultural policies and mismanagement of the schemes. This crisis not only hurts the schemes' populations, but contributes to the national food shortage. The schemes have failed to achieve the desired socioeconomic transformation and full involvement of their settlers. Moreover, most of the schemes have failed to recover the costs of establishing and even operating them.

Additionally, the state has greatly undermined the food security of the rural population by massively reducing their access to water, wood, land and pasture in favour of ecologically and environmentally unsound large-scale agricultural schemes. These modern schemes today control vast tracks of land and consume nearly all the production-enhancing inputs. In contrast, the traditional small-scale agricultural sector, which provides livelihoods and food for the majority of the rural population, controls little of the total land area under crops, and overuse has degraded much of that.

Agriculture in Sudan has failed to meet the needs of food security; hence the country has become a net importer of wheat. The production of sorghum and other grains are not enough to meet the population's food requirements. That situation is a result of interconnected and interrelated natural circumstances, such as fluctuating weather conditions, erratic rainfall and so on, and manmade reasons: lack of political stability; a top-down approach to development that has turned rural producers into policytakers rather than policymakers; weak government administrative and implementation capacity; and the low priority assigned to agriculture in the allocation resources.

Successive plans and strategies have been implemented to develop agriculture and reverse this failure, but with limited success in achieving their objectives. And the situation has been exacerbated since the turn of the millennium and the discovery of oil, as the government has invested in petroleum and related services at the expense of agriculture, which has been totally ignored and neglected.

Another dimension that negatively affects the agriculture sector in Sudan is the continuing influx of IDPs – because of climate change and recurrent droughts and desertification or because of wars and conflicts. These people, in relocating to cities and moving from one area to another, put a lot of strain on natural resources and contribute to land and environmental degradation. Increased temperatures and decreased precipitation because of global climate change will further worsen the situation in the agricultural sector and consequently the food situation.

Sustainable agricultural production in Sudan will depend on: the continued serviceability of agricultural machinery across the different schemes; availability of water and other inputs; dedicated and qualified management of scheme boards; regular maintenance of dams, canal systems and all related irrigation infrastructure; prompt payment for outputs; and continuing good collaboration between institutions and stakeholders.

Efficient production is possible in the future only if the government addresses the following critical issues:
1. Conflict resolution to prevent bloodshed and the loss of natural resources.
2. Timely provision of foreign exchange for imported annual inputs.
3. Provision of adequate operating and maintenance funding.
4. Continued improvement of the financial performance of schemes boards.
5. Retention of skilled managerial and technical staff.

References

Abbadi, K. A. and A. E. Ahmed (2006). Brief Overview of Sudan Economy And Future Prospects For Agricultural Development. *Khartoum Food Aid Forum*, The World Food Programme (WFP).

Abdalla, S. and M. K. (2010). Water Policy of Sudan: National And Co-Basin Approach, Ministry of Irrigation and Water Resources, Sudan. *Regional Centre on Urban Water Management*. Tehran.

Abdulatif, N. (2008). Irrigation Crisis Threatens Agricultural Season In New Halfa. *Sudan Vision Daily News paper official website*. Khartoum, Byader Media Distribution Co. Ltd: pp. 4.

AfDB, et al. (2012). African Economic Outlook 2012: Promoting Youth Unemployment. Issy les Moulineaux, OECD.

Ahmed, A. A. and U. H. Ismail (2008). Sediment in the Nile River System. Khartoum, UNESCO.

Ahmed, E., et al. (2012). "Interplay of Agriculture Deterioration and Poverty Incident." *International Journal of Agriculture and Forestry* 2(1): 18–29.

Ahmed, E., et al. (2012). "Mechanism of Poverty Incident in Agricultural Sector of Sudan." *Journal of Development and Agricultural Economics* 4(14): 371–383.

Ahmed, M. M. (2010). Sudan Phase 2. London, Overseas Development Institute (ODI). 19: 50.

Ahmed, S. M. and L. Ribbe (2011). "Analysis of Water Footprints of Rainfed and Irrigated Crops in Sudan." *Journal of Natural Resources and Development* 1(1):9.

Al-Naiem, N. (2009). "Economic: Gezira Scheme." *Sudan Vision*. Retrieved 28/03/2013, from http://www.sudanvisiondaily.com/modules.php?name=News&file=article&sid=46589.

Ali, A. A. (2010). Egypt's Takeover of Sudan's Gezira Scheme. Paris, SudanTribune 4.

Ali, H. M. (2012). "Unsustainability of Dams in the Sudan: The Present Situation of Resettled Halfawiyyin." *Sudan Knowledge*. Retrieved 30/03/2013, from http://www.sudanknowledge.org/index.php/component/remository/?func=fileinfo&id=109.

Ali, O. M. M. (2003). Environmental Impact Assessment From a Sudanese Perspective. Khartoum, UNEP.

Ashok, S. (2008). "Mission Not Yet Accomplished: Managing Water Resources In The Nile River Basin."*Journal of International Affairs (New York)* 61(2): 201–214.

Babiker, M. (1997). Resource Competition and the Future of Pastoralism in the Butana Plain. Khartoum, University of Khartoum.

Barton, A. and G. Writer (2012). "Water in Crisis – Spotlight On Sudan." *The Water Project*. Retrieved 28/03/2013, from http://thewaterproject.org/water-in-crisis-sudan.asp.

Bashir, M. (2001). National Biodiversity Strategy and Action Plan (NBSAP). Khartoum, UNDP. 1: 23.

Bernal, V. (1997). "Colonial Moral Economy and the Discipline of Development: The Gezira Scheme and "Modern" Sudan." *Cultural Anthropology* 12(4):447–479.

Bitsue, K. K. (2012). The Nile: From Mistrust and Sabre Rattling to Rapprochement. Addis Ababa, Institute for Security Studies. 238:12.

Boko, M., I. Niang, A. Nyong, C. Vogel, A. Githeko, M. Medany, B. Osman-Elasha, R. Tabo and P. Yanda, (2008). Climate Change 2007 : Impacts, Adaptation and Vulnerability: Working Group II Contribution to the Fourth Assessment Report of the IPCC Intergovernmental Panel on Climate Change. Geneva, IPCC

Boussard, J.-M., et al. (2005). "Food security and agricultural development in Sub-Saharan Africa: Building a case for more support." *Background document. Final Rep., FAO, Rome*: p.3.

Central Bureau of Statistics of Sudan, et al. (2010). Food And Nutrition Security Assessment In Sudan : Analysis of 2009 National Baseline Household Survey (NBHS). Khartoum, Central Bureau of Statistics.

Central Intellegence Agency, C. (2013, 09/03/2013). "Africa, Sudan." *The World Fact Book*. from https://www.cia.gov/library/publications/the-world-factbook/geos/su.html.

Central Intelligence Agency, C. (1991). "World Factbook, Sudan Soils ". Retrieved 15/03/2013, from http://www.photius.com/countries/sudan/geography/sudan_geography_soils.html.

Cesar Guvele, H. F., Eltahir Nur, Abdelaziz Abdelfattah and Aden Aw-Hassan, (2011). Poverty Assessment Southern Sudan. Aleppo, Syria, ICARDA.

Cox, T. P. (2010). "Conserving Water For Livestock In Butana, Sudan." *New Agriculturist*. Retrieved 30/03/2013, from http://www.new-ag.info/en/focus/focusItem.php?a=1675.

Dara Climate Vulnerable Forum (2012). Climate Vulnerability Monitor : A Guide to the Cold Calculus of a Hot Planet. Madrid, DARA:187–188.

Darwin, R. (2001). "Climate Change and Food Security." *Agriculture information bulletin*(765–8): 2.

Davies, H. R. J., Ed. (1985). Natural Resources and Rural Development in Arid Lands: Case Studies From Sudan. Tokyo, United Nations University.

Elageed, A. A. E. (2008). Weaving The Social Networks of Women Migrants in Sudan : The Case of Gezira. Münster, LIT Verlag.

Elasha, B.-O., et al. (2005). Sustainable Livelihood Approach for Assessing Community Resilience to Climate Change: Case Studies from Sudan, AIACC Working Paper.

Elasha, B. O. (2010). Mapping of Climate Change Threats and Human Development Impacts in the Arab Region. *UNDP Arab Development Report–Research Paper Series*. Cairo, UNDP Regional Bureau for the Arab States.

Eldaw, A. M. (2004). The Gezira Scheme: Perspectives for Sustainable Development. Bonn, German Development Institute. 2: 82.

Elgali, M. B., et al. (2010). Development of the Agricultural Crops Trade Sector of Sudan Under the Increasing World Food Prices. *2010 AAAE Third Conference/ AEASA 48th Conference*. Cape Town.

Ertsen, M. W. (2006). "Colonial Irrigation: Myths of Emptiness." *Landscape Research* 31(2): 147–167.

European Court of Auditors (2012). Effectiveness of European Union Development Aid for Food Security in Sub-Saharan Africa. Luxembourg. 1: 70.

Famine Early Warning Systems Network, F. (2012). Food Security Will Improve During Harvest And Post-Harvest Period. Khartoum: 7.

FAO/ SIFSIA. (2010). Cereal Availability Study in the Northern States of Sudan. Khartoum, FAO: 61.

FAO (2000). Wheat Production Potential in Sudan. Rome.

FAO (2005). Food Security and Agricultural Development in Sub-Saharan Africa, Building A Case For More Public Support. Rome.

FAO (2010). Plan of Action for North Sudan August 2010 – August 2012. Rome, Emergency Operations and Rehabilitation Division, FAO:122.

FAO (2012). "Cereal Supply and Demand Brief." *World Food Situation*. Retrieved 30/03/2013, from http://www.fao.org/worldfoodsituation/wfs-home/csdb/en/.

FAO (2012). Undernourishment around the World in 2012. Rome, FAO: 7.

FAO in emergency (2011). "The FAO Component Of The Consolidated Appeals 2012: Sudan." *Journal of Development and Agricultural Economics*. Retrieved 20/03/2013, from http://www.fao.org/emergencies/resources/documents/resources-detail/en/c/150065/.

FAO (2013). The State of Food Insecurity in the World. The multiple dimensions of food security. Rome, FAO: pp 1. Retrieved 20/12/2013, from http://www.fao.org/docrep/018/i3458e/i3458e.pdf

FAO, W. and IFAD (2012). The State of Food Insecurity in the World 2012. Economic growth is necessary but not sufficient to accelerate reduction of hunger and malnutrition. Rome, FAO:pp 8–11.

FAO/EWSFA (1997). Crop and Food Supply Situation in Sudan. Rome, FAO.

FAO/RNE (1998). Proceedings of the Second Expert Consultation on National Water Policy Reform in The Near East, Cairo, Egypt, FAO.

FAO/SIFSIA Programme (2005). Sudan Institutional Programme: Food Security Information For Action (Sifsia). Rome, FAO/SIFSIA Programme in northern Sudan.

Frenken, K. (2005). Irrigation in Africa in Figures: Aquastat Survey, 2005. Rome, FAO.

Gaitskell, A. (1959). Gezira: A Story of Development in the Sudan. London, Faber and Faber.

Government of Sudan (2003). Sudan's First National Communications under the United Nations Framework Convention on Climate Change. Khartoum, Higher Council for Environment and Natural Resources, HCENR. 1:111.

Government of Sudan (2007). National Adaptation Programme of Action (NAPA). Khartoum, Higher Council for Environment and Natural Resources.

Government of Sudan (2008). "About Sudan." Retrieved 07/03/2013, from http://www.sudan.gov.sd/en/index.php?option=com_content&id=50&Itemid=67.

Government of Sudan (2008). Excutive Summary of Executive Programme for Agricultural Revival. General Secretariat of Council of Ministers. Khartoum, Government of Sudan.

Graßl, H., et al. (2003). Climate Protection Strategies for the 21st Century: Kyoto and Beyond. Berlin, The German Advisory Council on Global Change (WBGU).

Gregory, P. J., et al. (2005). "Climate Change and Food Security." *Philosophical Transactions of the Royal Society B: Biological Sciences* 360(1463):2139–2148.

Hamid Faki, E. M. N., Abdelaziz Abdelfattah, Aden A Aw-Hassan, (2012). Poverty Assessment Northern Sudan. Aleppo, ICARDA:62.

Harrison, M. N. and J. K. Jackson (1958). Ecological Classification Of The Vegetation Of The Sudan. Khartoum, Agriculture Publications Committee.

Hassan, K. I. (1984). Production Relations in the Sudanese Agriculture: The Case of the Gezira Scheme. Khartoum, University of Khartoum, Faculty of Economic and Social Studies, Development Studies and Research Centre (DSRC).

Hassan, K. I. (2011). "The Gezira Scheme Role in Sustainable Development and Food Security (1)." Retrieved 05/06/2013, 2013, from http://www.sudanvisiondaily.com//modules.php?name=News&file=article&sid=75438

Himeidan, Y., et al. (2007). "Climatic Variables and Transmission of Falciparum Malaria in New Halfa, Eastern Sudan." *Eastern Mediterranean Health Journal* 13(1):17.

Holt, J. and L. Coulter (2011). Livelihoods Zoning "Plus" Activity in Sudan. Khartoum, The Famine Early Warning Systems Network (FEWS.NET).

Hornby, C. (2012). "UN Food Agencies: One In Eight People Are Going Hungry." Retrieved from http://www.reuters.com/article/2012/10/09/us-un-hungry-idUSBRE8980A220121009.

IFAD (1992). The Northern Province of Sudan Irrigation Rehabilitation Project. Khartoum.

IFAD (2009). Republic Of Sudan: Country Program Evaluation. Khartoum. 2060–SD:136.

IFAD (2010). "Rural Poverty in The Sudan." Retrieved 02/02/2013, from http://www.ruralpovertyportal.org/country/home/tags/sudan.

Institute for Security Studies (2005). "Sudan Economy Overview." *Knowledge empowers Africa*. Retrieved 22/03/2013 from http://www.issafrica.org/.

IRINAfrica (2007). "Climate Change – Only One Cause Among Many For Darfur Conflict " *Humanitarian news and analysis, a service of the UN Office for the Coordination of Humanitarian Affairs*. Retrieved 16/03/2013, from http://www.irinnews.org/Report/72985/SUDAN-Climate-change-only-one-cause-among-many-for-Darfur-conflict.

Ismail, S. " The Origins Of The Term "Nubia"." *Nubian civilization*. Retrieved 09/03/2013, from http://shazlyasmail.tripod.com/favorite.htm.

Johnson, D. H. (2003). The Root Causes of Sudan's Civil Wars. Bloomington, Indiana University Press.

Kimball, B., et al. (1993). "Effects of Increasing Atmospheric CO_2 On Vegetation." *Vegetatio* 104(1):65–75.

Konandreas, P. (2009). "Assessing Sudan's Export Diversification Potential in Agricultural Products." *Maxwell Stamp PLC for the Ministry of Foreign Trade, Government of Sudan and the European Commission*:117.

Lean, G. (2009). "Water Scarcity Now Bigger Threat Than Financial Crisis", By 2030, More Than Half The World's Population Will Live In High-Risk Areas. *The Independent*. Dublin, Independent News and Media. 15:2.

Maeng, M. H. (2012). "Climate Change, Food Security and Conflict." *Global health*. Retrieved 30/03/2013, from http://www.sipri.org/blogs/global-health/climate-change-food-security-and-conflict.

Mahir, M. E. A. E. and H. H. Abdelaziz (2010). "Analysis of the Effect of Price Liberalization Policy on Production of the Main Crops Grown in New Halfa Agricultural Corporation." *Research Journal of Agriculture and Biological Sciences* 6(6):878–884.

Mamdani, M. (2009). Saviors and Survivors : Darfur, Politics, and the War On Terror. New York, Pantheon Books.

McCartney, M. (2012). The Implications of Climate Change for Water Resource Development in the Blue Nile River, GWF, Discussion Paper 125, Global Water Forum. Canberra. 1251:5.

McCartney, M. P. and M. M. Girma (2012). "Evaluating the Downstream Implications of Planned Water Resource Development in the Ethiopian Portion of the Blue Nile River." *Water International* 37(4):362–379.

Metz, H. C. (1992). Sudan : A Country Study. Washington, D.C., Federal Research Division, Library of Congress.

Moges, S. A., et al. (2011). Climate Change and Water in Africa : Analysis Of Knowledge Gaps And Needs. Addis Ababa, United Nations Economic Commission for Africa (UNECA). 4:23.

Mohamed-Ali, S., et al. (2009). Water Resources in Sudan: Enhancing Rainfall Harvesting Methods for Water Supply. *World Environmental and Water Resources Congress 2009: Great Rivers* Kansas City, Missouri, American Society of Civil Engineers

Mohamed Alameen, E. E. A. (2009). The Reality of Execution of Some Agricultural Extension Policies: Case Study Gezira State, Sudan. *Faculty of Agricultural Sciences*. Wad-Madani, University of Gezira.

Mongabay (1991). "Sudan-Agriculture, Livestock, Fisheries and Forestry." Retrieved 30/01/2013 from http://www.mongabay.com/history/sudan/sudan-agriculture,_livestock,_fisheries,_and_forestry.htm.

Mwaniki, A. (2006). "Achieving Food Security in Africa: Challenges and Issues." *UN Office of the Special Advisor on Africa (OSAA) http://www. (Last accessed on 09/05/2010)*.

Netherlands Water Partnership (2007). Smart Water Harvesting Solutions: Examples of Innovative, Low-Cost Technologies for Rain, Fog, Runoff Water and Groundwater. Hague, Netherlands Water Partnership (NWP). 1:65.

Nimir, M. B. and I. A. Elgizouli (2011). Climate Change Adaptation and Decision Making in the Sudan. Washington, DC, World Resources Institute.

Omer, A. M. (2011). "Agriculture Policy in Sudan." *Agricultural Science Research Journal* Vol 1(1): pp.1–29.

Ongley, E. D. (1996). Control of Water Pollution from Agriculture. Rome, FAO.

Pimbert, M. (2009). Towards Food Sovereignty. London, International Institute for Environment and Development: pp.2–9.

Practical Action (2012). "Increasing Resilience To Poverty In Kassala, '99,000 Reasons to Smile." Retrieved 30/03/2013 from http://practicalaction.org/kassala-news.

Ravallion, M. (2009). A Comparative Perspective on Poverty Reduction in Brazil, China and India. *World Bank Policy Research Working Paper Series*. Washington DC, World Bank. 5080:39.

Reid, H. (2009). Community-Based Adaptation to Climate Change. London, International Institute for Environment and Development.

Ringler, C., et al. (2010). Climate Change Impacts on Food Security in Sub-Saharan Africa: Insights From Comprehensive Climate Change Scenarios. Addis Ababa, International Food Policy Research Institute (IFPRI). 01042:12–14.

Rosset, P. (2003). "Food sovereignty: global rallying cry of farmer movements." *Backgrounder* 9(4):1–4.

Salman, S. M. (2011). "The New State of South Sudan and the Hydro-Politics Of The Nile Basin." *Water International* 36(2):154–166.

Salman, S. M. A. (2010). "Water Resources in the Sudan North-South Peace Process: Past Experience and Future Trends." *African Yearbook of International Law* 16:299–328.

Salman, S. M. A. (2013). The Gezera Scheme: Historical Background and Developments- Promulgated in Arabic. *Algrar*. Khartoum, Algrar Company.

Salman, S. M. A. (2013). "Sudan has failed to use 350 billion cubic meters of its share of the Nile water." *Sudanile*. Retrieved 27/06/2013, from http://www.sudanile.com/index.php?option=com_content&view=article&id=53963:------350--------&catid=42:2008-05-19-17-16-29&Itemid=60.

Schmidhuber, J. and F. N. Tubiello (2007). "Global Food Security Under Climate Change." *Proceedings of the National Academy of Sciences* 104(50):19703–19708.

Sheehy, J. E., et al. (2008). Harnessing Photosynthesis in Tomorrow's World: Humans, Crop Production And Poverty Alleviation. *Photosynthesis. Energy from the sun*, Springer:1237–1242.

Shinn, D. H. (2006). "Nile Basin Relations: Egypt, Sudan And Ethiopia." Retrieved 18/02/2013, from http://elliott.gwu.edu/news/speeches/shinn0706_nilebasin.cfm.

Shukri Ahmed, et al. (2007). FAO/WFP Crop and Food Supply Assessment Mission to Sudan Special Report. Rome, FAO.

Siddig, K. H. and B. I. Babiker (2012). "Agricultural Efficiency Gains and Trade Liberalization In Sudan." *AfJARE* 7(1).

Stads, G. and K. ElSiddig (2010). Recent Developments in Agricultural Research. Rome, FAO.

Strzepek, K. and A. McCluskey (2007). The Impacts of Climate Change on Regional Water Resources and Agriculture in Africa. *World Bank Policy Research Working Paper*. Washington D C, World Bank. 4290:68.

Sudan and Egypt (1959). "No. 6519. Agreement Between the Republic of the Sudan and the United Arab Republic for the Full Utilization of the Nile Waters. Signed at Cairo, on 8 November 1959." *The Program in Water Conflict Management and Transformation (PWCMT)*. Retrieved 12–09, 2013, from http://ocid.nacse.org/tfdd/tfdddocs/230ENG.pdf.

Sudan Embassy, S. (2011). "General Information." Retrieved 09/03/2013, from http://sudanembassy.se/?p=88.

Sullivan, D. P. J. (2010). Perspective: Sudan – Land of Water and Thirst; War and Peace. Traverse City, MI, Circle of Blue Organization.

Sørbø, G. M. (1985). Tenants and Nomads in Eastern Sudan: A Study of Economic Adapations in the New Halfa Scheme. Motala, Motala Grafiska AB.

Taha, F. (2010). The History of the Nile Waters in the Sudan. *The River Nile in the post-colonial age*. T. Tvedt. London, IB Tauris:179–216.

Tamiotti, L., et al. (2009). Trade And Climate Change : A Report by the United Nations Environment Programme and the World Trade Organization. Geneva, World Trade Organization.

The Water Project (2012). Improving Health in Africa Begins with Access to Safe Water. Concord, The Water Project Organization.

Tvedt, T. (2004). The River Nile in the Age of the British [Elektronisk Resurs] Political Ecology and the Quest For Economic Power. London, I.B. Tauris.

U. S. Department of State (2012). "Sudan." Retrieved 28/03/2013, from http://www.state.gov/outofdate/bgn/sudan/194934.htm.

UNDP (2006). Macroeconomic Policies for Poverty Reduction: The Case of Sudan. Khartoum, United Nations Development Programme (UNDP).

UNDP (2006). Pastoral Production Systems in South Kordofan. Khartoum. 2:30.

UNDP (2012). Sudan: Human Development Report 2013:4.

UNEP (2007). Sudan : Post-Conflict Environmental Assessment. Nairobi, United Nations Environment Programme.

UNICEF (2004). Towards A Baseline: Best Estimates of Social Indicators for Southern Sudan. Nairobi, New Sudan Centre for Statistics and Evaluation, in association with UNICEF. NSCSE series paper. 1:74.

UNICEF (2010). Unicef In Sudan Marks World Water Day 2010 With Focus On Water Quality. Khartoum, UNICEF.

Urama, K. C. and N. Ozor (2010). "Impacts of Climate Change on Water Resources in Africa: The Role Of Adaptation." *African Technology Policy Studies Network*: 29.

USAID (2012). "Heavy Rains Trigger Widespread Flash Flooding." Retrieved 1803/2013, from http://reliefweb.int/sites/reliefweb.int/files/resources/Sudan_FSOU_2012_08_final.pdf

USAID and F. NET (2012). "Food Prices Continue To Increase In Most Reference Markets And Remain Above Average." Retrieved 20/03/2013, from http://www.fews.net/docs/Publications/Sudan_FSOU_2012_06_final.pdf.

USAID and F. NET (2012). "Significant Food Consumption Gaps Expected Across Border Areas Of Sudan And South Sudan ". Retrieved 20/03/2013, from http://www.fews.net/docs/Publications/Sudan_South_Sudan_Alert_2012_07_final.pdf.

USAID, F. N. (2012). "Intensified Conflict In Border Areas Increases The Size Of The Food Insecure Population." Retrieved 30/03/2013, from http://www.fews.net/contact/Pages/contact.aspx?l=en.

Wallach, B. (1988). "Irrigation In Sudan Since Independence." *The Geographical Review* 74(2):417–434.

Watch, G. (2011). "South Sudan's Country Profile: The Birth of A New Nation." *Genocide Watch, The International Alliance to End Genocide*. Retrieved 15/01/2013, from http://www.genocidewatch.org/southsudan.html.

Webb, P. and B. L. Rogers (2003). Addressing The" In" In Food Insecurity. Washington, DC, Food and Nutrition Technical Assistance Project Academy for Educational Development.

WHO (2009). "10 Facts About Water Scarcity." Retrieved 28/03/2013, from http://www.who.int/features/factfiles/water/en/index.html.

Von Braun, J. (2007). The World Food Situation: New Driving Forces And Required Actions. Washington,D.C., Intl Food Policy Res Inst.

World Bank (1978). Sudan : New Halfa Irrigation Rehabilitation Project. Washington, D.C., World Bank: 81.

World Bank (2000). Sudan : Options for the Sustainable Development of the Gezira Scheme. Washington, D.C., World Bank.

World Bank (2009). "Agriculture: An Engine for Growth and Poverty Reduction." *IDA at Work*. 2013, from http://siteresources.worldbank.org/IDA/Resources/IDA-Agriculture.pdf.

World Bank (2012). "Sudan." Retrieved 03/09/2013 from http://data.worldbank.org/country/sudan.

World Food Summit and FAO (1996). Rome Declaration on World Food Security and World Food Summit Plan of Action. *World Food Summit*, Rome, FAO.

Zaroug, M. G. (2006). Country Pasture/Forage Resource Profiles. Rome, FAO.

Zuberi, T. and K. J. Thomas (2012). Demographic Projections, The Environment And Food Security In Sub-Saharan Africa. 1:24.

Appendix

In the Name of God, The Merciful, The Compassionate

THE GEZIRA SCHEME ACT of 2005

Pursuant to the provisions of the Constitution of the Republic of the Sudan of 1998, the National Council hereby has adopted, and the President has signed the following Act:

Chapter One
Preliminary Provisions
Citation of the Act and Effective Date

1. This Act shall be cited as "The Gezira Scheme Act of 2005" and shall come into effect upon signature.

Rescission

2. (1) The Gezira Scheme Act of 1984 is hereby repealed but all bylaws, orders and regulations issued thereunder shall remain valid and applicable until repealed or amended in accordance with the provisions of this Act.

   (2) The Gezira Land Act of 1927 is hereby repealed but all bylaws, orders and regulations issued thereunder shall remain in valid and applicable until repealed or amended in accordance with the provisions of this Act.

Interpretation

3. In this Act, unless the context may otherwise require, the following words and phrases shall have the meanings hereby assigned to them:
   Government: Refers to the Federal Government.
   Competent Minister: Refers to the Federal Minister of Agriculture and Forestry
   Scheme: Refers to the Gezira Scheme with its current command areas or any extension thereof.
   Board: Refers to the Scheme's Board of Directors established pursuant to Article (6) hereof.
   General Manager: Refers to the Scheme's General Manager who is appointed in accordance with Article (15) hereof.
   Farmer: Refers to any individual with a hawasha (lot) in accordance with Article (16) hereof.
   Employee: Refers to any individual hired within the Scheme's job structure.

Fiscal Year: Refers to the period of twelve months commencing on July 1 of each year and ending on June 30 of the following year, or any date to be set by the Board of Directors for the beginning and end of the fiscal year.

Field Canals: Refer to irrigation canals called secondary canals, "Abou Ishreenat" and "Abou Sittat" and to water regulators and controllers.

Irrigation Canals: Refer to primary and secondary canals and to major canals and drainages, including regulators and pipes branching out to feed field canals.

Water User Association: Refers to farmers organizations undertaking actual tasks with regard to water management, operation and uses.

Chapter Two
Scheme's Identity, Headquarters and Sponsorship

4. (1) The Gezira Scheme, which is an economic and social entity with various activities and enjoys national support as part of the development effort, shall be established pursuant to this Act. It shall be an administratively, financially and technically independent juridical person with a permanent, perpetual character and a public seal. It shall have the right to litigate in its own name.

   (2) The State, as represented by the Ministry of Finance and National Economy, shall own the current assets of this Scheme, and may allow future participation of private sector investments, whether in current assets or through addition of new assets to the Scheme.

   (3) The Gezira Scheme shall be composed of:

   A Farmers
   B The Government, as represented by its respective units providing basic services such as development, irrigation and public goods, including: research, plant protection, technology support, agricultural extension, technical studies, training as well as supervisory management and indicative planning.
   C Private sector with regard to provision of auxiliary commercial services.

   (4) The Scheme shall be headquartered in the City of Barakat. The Board of Directors may create branches and offices in or outside the Sudan whenever necessary.

   (5) The Scheme shall be under the aegis of the competent Minister.

Scheme's Objectives

5. The Scheme aims to utilize its sustainable and stable agricultural production resources and potentials to improve the economic and social standards of farmers and its own employees as well as the area it is located in, and to improve respective services provision. It also aims to contribute to the attainment of national objectives. Without prejudice to the foregoing, the Scheme shall aim to:

    (a) Achieve optimal and rational utilization of the Scheme's resources and potentials to increase income level, boost agricultural output and maximize benefits and returns.
    (b) Achieve the Scheme's local and national objectives, such as food security, job creation, increased and diversified exports, and introduction of manufacturing industries.
    (c) Achieve citizen's well-being within the Scheme through economic development.
    (d) Preserve the environment within the boundaries of the Scheme.
    (e) Ensure farmers' right to freely manage their production and economic aspects within the technical parameters, and employ technology support to boost production and maximize their respective returns.
    (f) Ensure farmers' right to effectively participate, at all administrative levels, in planning and implementation of projects and programs that affect their production and livelihoods
    (g) Ensure farmers' right to manage irrigation operations at field canal level through water user associations.
    (h) Promote farmers' effective collective action to ensure efficient provision of services and economic production while maximizing economies of scale.
    (i) Provide an opportunity to the private sector to play a leading role in provision of auxiliary commercial services in a competitive environment.
    (j) Introduce irrigated forestry and livestock in agricultural cycle.
    (k) Provide auxiliary services to the Scheme's activities by competent authorities.

Chapter Three
Board of Directors
Board Composition

6   (1) A Board of Directors shall be established by a Decree of the President of the Republic, upon a recommendation by the competent Minister, and shall be composed of a chairman and 14 members, as follows:

a. A Board chairman appointed by the President.
  b. The General Manager in ex officio capacity.
  c. Farmers' Union representatives who shall form at least 40 % of the Board membership.
  d. A representative from the Scheme's employees.
  e. Representatives from the relevant Ministries.

(2) The positions of Board chairman and General Manager shall not be held by one and the same person.

Conditions for Board Membership
7. A member of the Board of Directors must be:
   a. A Sudanese citizen who is mentally competent.
   b. A person who has not been convicted of a crime of moral turpitude or a breach of honor.
   c. A person who has not been declared bankrupt.
   d. A literate person who meets all eligibility requirements

Declaring a Vacancy and Appointing a Replacement
8. (1) The Board position of any member shall become vacant for any of the following reasons:
   a. Loss of any of the conditions of membership stipulated in Article (7) of this Act.
   b. Resignation.
   c. Being relieved of duties or removed by the appointing authority.
   d. Death of the incumbent.
   e. Failure to attend three successive meetings without an acceptable justification.

(2) In the event that a Board membership position has become vacant, a replacement shall be appointed following the same procedures applied in appointing the replaced member.

Board Functions and Powers
9. The Board shall be entrusted with formulating plans and general policies for achieving the Scheme's objectives. Without prejudice to the foregoing, the Board shall have the following functions and powers:
   a. Developing scientific parameters for research, economic and social studies required to ensure optimal utilization of the Scheme's resources to achieve the highest possible profit rates.
   b. Developing equitable incentive policies in order to carry out the State's strategic policies with regard to agricultural produces.

c. Managing and developing basic services of research, plant protection, technology support, agricultural extension, seeds multiplication, training and inner roads.
d. Establishing a burden-sharing [social safety] system allowing compensations for earnest farmers in the event of exposure to pests and natural catastrophes.
e. Establishing technical parameters for cropping patterns and agricultural cycles.
f. Approving plans and programs submitted by the General Manager.
g. Determining its charges categories of services performed in coordination and agreement with competent authorities, and respective charges to be levied on farmers by water user associations.
h. Approving employment of workers whom the Board may deem necessary to carry out its functions in accordance with the job structure it approved, and issuing regulations governing their employment.
i. Approving the annual draft budget for running the Scheme, as well as the development budget to be discussed with the Minister of Finance and National Economy to determine the required development support.
j. Ensuring optimal use of the Scheme's assets and of moveable and immoveable property owned by the government, and investing the same economically.
k. Maintaining the Scheme's lands and taking measures necessary to preserve the soil.
l. Providing information to help farmers market their produces.
m. Entering into contracts and agreements necessary to carry out the Scheme's objectives.
n. Carrying out other actions as may be deemed necessary or supportive by the Board to realize the Scheme's objectives.
o. Forming permanent or temporary committees to assist the Board of directors.
p. Delegating any of its functions or powers to the General Manager or any of its committees.
q. Issuing the necessary by-laws to regulate its activities.

Meetings of the Board

10. (1) The Board shall hold at least six meetings during a fiscal year. The Board Chairman may call an extraordinary meeting when necessary or at the written request of half of the Board members.

(2) The quorum for Board meetings shall be met with the simple majority of the members.
(3) Board resolutions shall be adopted by majority of votes of the members present, and the Board chairman shall have the casting vote in the event of a tie.
(4) The Board chairman may ask any Board member to chair the meeting in his absence.
(5) The Board may invite any person to attend any of the Board or committee meetings but the invitee shall not have the right to vote.
(6) All Board procedures and discussions shall be confidential and no member shall disclose any information about them before they are made public in the manner to be decided by the Board.

Disclosure of Interest
11. Any member of the Board or affiliated committees who has a vested direct or indirect interest in any matter, proposal or topic submitted to the Board or any of its committees for discussion must disclose this interest to the Board or committee. He shall not participate in any discussion or resolution adopted by the Board or committee in this respect.

Remuneration of Board and Committees
12. The Board shall determine the remuneration package of the Board chairman, members and affiliated committees in accordance with financial regulations.

General Secretariat
13. The Board shall have a general secretariat, headed by a Secretary appointed by the Board. The Board shall specify his terms of reference and duties.

General Manager
14. (1) The Board shall enter into a contract with a highly competent and experienced person to be the Scheme's General Manager.
(2) Said contract shall specify a renewable four-year-term as well as terms of employment and remuneration of the General Manager.

Powers and Duties of the General Manager
15. The General Manager shall be the most senior executive officer charged with implementation of the Board's policies, plans and programs. Without prejudice to the above, the General Manager shall oversee and have powers in respect of the following:

a. Provision of agricultural extension services to enable farmers to apply appropriate technology to increase production and productivity.
b. Taking measures necessary to provide protective services for crops against pests and diseases.
c. Taking measures necessary to provide seeds multiplication services.
d. Carrying out the Scheme's sustainable economic development with regard to its services.
e. Preparing the annual budget and submitting the same to the Board in a timely fashion before the commencement of the fiscal year.
f. Preparing the Annual Report, listing what has been accomplished in the previous year and including performance indicators for the following year. The Report shall be submitted to the Board of directors at least one month before commencement of the new fiscal year.
g. Monitoring the progress of work in the Scheme's facilities and submitting periodic reports to the Board in a timely fashion.
h. Submitting to the Board recommendations for the appointment and promotion of the Scheme's employees in accordance with the regulations issued by the Board.
i. Disciplining of employees and transferring them in accordance with regulations issued by the Board.
j. Ruling on grievances submitted by employees in accordance with applicable by-laws.
k. The General Manager may delegate any of his powers to any of his assistants.

Chapter Four
Ownership of Hawashas, Irrigation and Drainage
Ownership of Hawashas
16. (1) All hawashas allocated to the farmers in the Scheme prior to the issuance of this Act shall be considered as though they had been allocated pursuant to the provisions of this Act.
    (2) The government shall take the following necessary measures:
    a. Farmers, holding land in freehold, to whom hawashas are allocated under such ownership, shall be allowed to have such hawashas registered in their names as freehold in the Directorate of Land Registration.
    b. Farmers who have not been allocated any hawashas during distribution and who have surplus land according to para (a), their land title shall be transferred to the Scheme and they shall be fairly compensated thereof.

c. The rest of the farmers who do not have freehold hawashas in the Scheme shall have the hawashas they currently possess registered in their names as leasehold for a period of ninety nine years.

(3) The new owners of the Hawashas shall pay charges to be determined by the Board in return for the freehold title registration.

(4) The Board shall have the right to issue regulations governing optimal utilization of hawashas in accordance with public agricultural policies, as well as other regulations necessary for implementation of technical controls for owners.

(5) Utilization of the Hawashas shall be governed by the following conditions:
  a. Using Hawashas strictly for agricultural purposes.
  b. No fragmentation of landholding.
  c. Sale or transfer of ownership shall be governed by the right of acquisition by pre-emption.

Disposing of Hawashas

17. (1) Subject to para. (5) C of Article 16, a farmer may dispose of his hawasha by sale, mortgage or assignment in accordance with directions set by the Board.
    (2) The Board shall have the right to determine the minimum size for ownership of hawasha.

Irrigation and Drainage

18. (1) The Ministry of Irrigation and Water Resources shall be responsible for operation and management of the primary irrigation and drainage canals and pumps in the Scheme, and for providing sufficient water for water user associations at the mouth of the respective field canals, and the Ministry of Finance and National Economy shall be responsible for financing maintenance, rehabilitation and operations of water canals in return for water charges to ensure provision of such services.
    (2) Water user associations shall maintain, operate and manage field canals and internal drainage.
    (3) All irrigation operations for any part within the Scheme command area shall have to be approved by the Board.

Water User Associations

19. (a) Water user associations shall be established under supervision of the Board at the Scheme level. They shall be legal entities representing the farmers' self-management system. They shall also undertake

actual responsibilities in managing water uses through entering into a contract with the Ministry of Irrigation and Water Resources in the area of supply of water and technical consultation.

(b) The Ministry of Irrigation and Water Resources shall establish a separate department for Gezira Scheme irrigation.

Chapter Five
Financial Provisions
Vesting of Assets and Rights

20. (1) To the Scheme shall be vested with:
    a. All assets and rights that had been transferred to Gezira Scheme pursuant to Gezira Scheme Act 1984.
    b. All debts and liabilities due from Gezira Scheme pursuant to Gezira Scheme Act 1984.
  (2) a. The assets, rights, debts and liabilities that have been transferred to the Scheme shall be assessed pursuant to Item (1) above, and the net assessed value shall be entered in the records of the Scheme.
    b. The Board may take any measures necessary for privatization of the cost centers.

Capital of the Scheme

21. The capital of the Scheme shall consist of the following:
    a. Accruals to the Scheme pursuant to the provisions of Article (20) of this Act.
    b. Allocations earmarked by the State for the Scheme.
    c. Funds and charges earned by the Scheme as a result of its activities or in return for services provided and privileges and exemptions granted to it.
    d. Grants and technical assistance accepted by the Board of Directors.
    e. Any other legitimate resources approved by the Board of directors.

Utilization of Scheme Resources

22. The Scheme financial resources shall be used to achieve its objectives. Without prejudice to the above, these financial resources shall be used as follows:
    a. Managing the Scheme and executing its activities, plans and programs.
    b. Paying the Scheme's financial obligations.
    c. Defraying the Scheme's expenses, including depreciation and replacement.

d. Paying employees' salaries, wages, bonuses, benefits and allowances and retirement benefits as well as remunerations to Board chairman and members.

Budget of the Scheme

23. 1) The Scheme shall have a separate operating budget prepared according to sound accounting principles issued by the Board of directors.
    (2) The General Manager shall prepare the development and rehabilitation budget and forward the same to the Board of directors for discussion and approval. He shall then submit the same to the Ministry of Finance to attain respective support for the budget's various components, including irrigation, research and technology support under the mandate given to the Board of Directors.
    (3) The Board of Directors shall approve the draft annual budget.
    (4) Surplus budget shall be used to develop and improve the Scheme.

Opening Bank Accounts and Maintaining Asset Records

24. a. The Board shall select the banks where the Scheme's bank accounts shall be opened in local and foreign currencies.
    b. The General Manager shall determine the persons authorized to handle those accounts.
    c. The Scheme shall maintain a regular record of fixed assets, which shall be audited annually.

Depreciation and Replacement Account

25. (1) The Scheme shall maintain a separate account for depreciation and replacement. It shall only be used for the purposes specified therein.
    (2) The Board may write off the value of obsolete assets by subtracting the items so designated from the depreciation and replacement account.

Accounts and Audit

26. (1) The Scheme shall maintain proper and accurate accounts in accordance with proper accounting principles.
    (2) The Auditor General or any auditor(s) approved by the Auditor General shall audit the Scheme's accounts at the end of each fiscal year.

Balance Sheet and Statements

27. The General Manager shall submit to the Board within three months from the close of the fiscal year a final account statement and the auditor's report on the Scheme's accounts.

Chapter Six
Transitional Provisions
28. (1) The Scheme's workers shall continue to carry out their mission until the job structure is approved and the terms of their services and employment are determined.
    (2) The cost centers at the Scheme shall continue to operate until they are privatized.
    (3) Farmers shall continue to hold Hawashas in the Scheme when this Act is adopted until the provisions specified in Article (16) hereof are implemented.
    (4) Responsibility for field canals shall be passed, after rehabilitation, to the water user associations.

Chapter Seven
Final Provisions
Primacy of the Provisions of this Act
29. In case of contradictions with any other law, the provisions of this Act shall prevail to the extent that such contradiction is removed.

Authority to Issue Regulations
30. The Board of directors may issue necessary bi-laws and regulations to implement the provisions of this Act.

IN WITNESS WHEREOF, the National Council has hereby passed "GEZIRA SCHEME ACT of "2005" in its meeting No. (18) of its 9th session dated 16 Jumada Al-Awal 1426 H. corresponding to June 22, 2005.

Ahmed Ibrahim El-Tahir
Speaker of the National Council

This Act was assented to by the President of the Republic on 6 July 2005.

CURRENT AFRICAN ISSUES PUBLISHED BY THE INSTITUTE

Recent issues in the series are available electronically
for download free of charge www.nai.uu.se

1981

1. *South Africa, the West and the Frontline States. Report from a Seminar.*
2. Maja Naur, *Social and Organisational Change in Libya.*
3. *Peasants and Agricultural Production in Africa. A Nordic Research Seminar. Follow-up Reports and Discussions.*

1985

4. Ray Bush & S. Kibble, *Destabilisation in Southern Africa, an Overview.*
5. Bertil Egerö, *Mozambique and the Southern African Struggle for Liberation.*

1986

6. Carol B.Thompson, *Regional Economic Polic under Crisis Condition. Southern African Development.*

1989

7. Inge Tvedten, *The War in Angola, Internal Conditions for Peace and Recovery.*
8. Patrick Wilmot, *Nigeria's Southern Africa Policy 1960–1988.*

1990

9. Jonathan Baker, *Perestroika for Ethiopia: In Search of the End of the Rainbow?*
10. Horace Campbell, *The Siege of Cuito Cuanavale.*

1991

11. Maria Bongartz, *The Civil War in Somalia. Its genesis and dynamics.*
12. Shadrack B.O. Gutto, *Human and People's Rights in Africa. Myths, Realities and Prospects.*
13. Said Chikhi, Algeria. *From Mass Rebellion to Workers' Protest.*
14. Bertil Odén, *Namibia's Economic Links to South Africa.*

1992

15. Cervenka Zdenek, *African National Congress Meets Eastern Europe. A Dialogue on Common Experiences.*

1993

16. Diallo Garba, *Mauritania–The Other Apartheid?*

1994

17. Zdenek Cervenka and Colin Legum, *Can National Dialogue Break the Power of Terror in Burundi?*
18. Erik Nordberg and Uno Winblad, *Urban Environmental Health and Hygiene in Sub-Saharan Africa.*

1996

19. Chris Dunton and Mai Palmberg, *Human Rights and Homosexuality in Southern Africa.*

1998

20. Georges Nzongola-Ntalaja, *From Zaire to the Democratic Republic of the Congo.*

1999

21. Filip Reyntjens, *Talking or Fighting? Political Evolution in Rwanda and Burundi, 1998–1999.*
22. Herbert Weiss, *War and Peace in the Democratic Republic of the Congo.*

2000

23. Filip Reyntjens, *Small States in an Unstable Region – Rwanda and Burundi, 1999–2000.*

2001

24. Filip Reyntjens, *Again at the Crossroads: Rwanda and Burundi, 2000–2001.*
25. Henning Melber, *The New African Initiative and the African Union. A Preliminary Assessment and Documentation.*

2003

26. Dahilon Yassin Mohamoda, *Nile Basin Cooperation. A Review of the Literature.*

2004

27. Henning Melber (ed.), *Media, Public Discourse and Political Contestation in Zimbabwe.*

28. Georges Nzongola-Ntalaja, *From Zaire to the Democratic Republic of the Congo.* (Second and Revised Edition)

2005

29. Henning Melber (ed.), *Trade, Development, Cooperation – What Future for Africa?*
30. Kaniye S.A. Ebeku, *The Succession of Faure Gnassingbe to the Togolese Presidency – An International Law Perspective.*
31. J.V. Lazarus, C. Christiansen, L. Rosendal Østergaard, L.A. Richey, Models for Life – Advancing antiretroviral therapy in sub-Saharan Africa.

2006

32. Charles Manga Fombad & Zein Kebonang, *AU, NEPAD and the APRM – Democratisation Efforts Explored.* (Ed. H. Melber.)
33. P.P. Leite, C. Olsson, M. Schöldtz, T. Shelley, P. Wrange, H. Corell and K. Scheele, *The Western Sahara Conflict – The Role of Natural Resources in Decolonization.* (Ed. Claes Olsson)

2007

34. Jassey, Katja and Stella Nyanzi, *How to Be a "Proper" Woman in the Times of HIV and AIDS.*
35. M. Lee, H. Melber, S. Naidu and I. Taylor, *China in Africa.* (Compiled by Henning Melber)
36. Nathaniel King, *Conflict as Integration. Youth Aspiration to Personhood in the Teleology of Sierra Leone's 'Senseless War'.*

2008

37. Aderanti Adepoju, *Migration in sub-Saharan Africa.*
38. Bo Malmberg, *Demography and the development potential of sub-Saharan Africa.*
39. Johan Holmberg, *Natural resources in sub-Saharan Africa: Assets and vulnerabilities.*
40. Arne Bigsten and Dick Durevall, *The African economy and its role in the world economy.*
41. Fantu Cheru, *Africa's development in the 21st century: Reshaping the research agenda.*

2009

42. Dan Kuwali, *Persuasive Prevention. Towards a Principle for Implementing Article 4(h) and R2P by the African Union.*
43. Daniel Volman, *China, India, Russia and the United States. The Scramble for African Oil and the Militarization of the Continent.*

2010

44. Mats Hårsmar, *Understanding Poverty in Africa? A Navigation through Disputed Concepts, Data and Terrains.*

2011

45. Sam Maghimbi, Razack B. Lokina and Mathew A. Senga, *The Agrarian Question in Tanzania? A State of the Art Paper.*
46. William Minter, *African Migration, Global Inequalities, and Human Rights. Connecting the Dots.*
47. Musa Abutudu and Dauda Garuba, *Natural Resource Governance and EITI Implementation in Nigeria.*
48. Ilda Lindell, *Transnational Activism Networks and Gendered Gatekeeping. Negotiating Gender in an African Association of Informal Workers.*

2012

49. Terje Oestigaard, *Water Scarcity and Food Security along the Nile. Politics, population increase and climate change.*
50. David Ross Olanya, *From Global Land Grabbing for Biofuels to Acquisitions of AfricanWater for Commercial Agriculture.*

2013

51. Gessesse Dessie, *Favouring a Demonised Plant. Khat and Ethiopian smallholder enterprise.*

52. Boima Tucker, *Musical Violence. Gangsta Rap and Politics in Sierra Leone.*
53. David Nilsson, *Sweden-Norway at the Berlin Conference 1884–85. History, national identity-making and Sweden's relations with Africa.*
54. Pamela K. Mbabazi, *The Oil Industry in Uganda; A Blessing in Disquise or an all Too Familiar Curse? Paper presented at the Claude Ake Memorial Lecture.*
55. Måns Fellesson & Paula Mählck, *Academics on the Move. Mobility and Institutional Change in the Swedish Development Support to Research Capacity Buildiing in Mozambique.*
56. Clementina Amankwaah. *Election-Related Violence: The Case of Ghana.*
57. Farida Mahgoub. *Current Status of Agriculture and Future Challenges in Sudan.*

www.ingramcontent.com/pod-product-compliance
Ingram Content Group UK Ltd.
Pitfield, Milton Keynes, MK11 3LW, UK
UKHW051652180426
11947UKWH00021B/1916

9 789171 067487